RF CMOS Oscillators for Modern Wireless Applications

RIVER PUBLISHERS SERIES IN CIRCUITS AND SYSTEMS

Series Editors:

MASSIMO ALIOTO
National University of Singapore
Singapore

KOFI MAKINWA
Delft University of Technology
The Netherlands

DENNIS SYLVESTER
University of Michigan
USA

Indexing: All books published in this series are submitted to the Web of Science Book Citation Index (BkCI), to SCOPUS, to CrossRef and to Google Scholar for evaluation and indexing.

The "River Publishers Series in Circuits & Systems" is a series of comprehensive academic and professional books which focus on theory and applications of Circuit and Systems. This includes analog and digital integrated circuits, memory technologies, system-on-chip and processor design. The series also includes books on electronic design automation and design methodology, as well as computer aided design tools.

Books published in the series include research monographs, edited volumes, handbooks and textbooks. The books provide professionals, researchers, educators, and advanced students in the field with an invaluable insight into the latest research and developments.

Topics covered in the series include, but are by no means restricted to the following:

- Analog Integrated Circuits
- Digital Integrated Circuits
- Data Converters
- Processor Architecures
- System-on-Chip
- Memory Design
- Electronic Design Automation

For a list of other books in this series, visit www.riverpublishers.com

RF CMOS Oscillators for Modern Wireless Applications

Masoud Babaie

Delft University of Technology
The Netherlands

Mina Shahmohammadi

Catena
The Netherlands

Robert Bogdan Staszewski

University College Dublin
Ireland

LONDON AND NEW YORK

Published 2019 by River Publishers
River Publishers
Alsbjergvej 10, 9260 Gistrup, Denmark
www.riverpublishers.com

Distributed exclusively by Routledge
4 Park Square, Milton Park, Abingdon, Oxon OX14 4RN
605 Third Avenue, New York, NY 10158

First published in paperback 2024

RF CMOS Oscillators for Modern Wireless Applications / by Masoud Babaie, Mina Shahmohammadi, Robert Bogdan Staszewski.

Routledge is an imprint of the Taylor & Francis Group, an informa business

Publisher's Note
The publisher has gone to great lengths to ensure the quality of this reprint but points out that some imperfections in the original copies may be apparent.

While every effort is made to provide dependable information, the publisher, authors, and editors cannot be held responsible for any errors or omissions.

ISBN: 978-87-93609-49-5 (hbk)
ISBN: 978-87-7004-351-9 (pbk)
ISBN: 978-1-003-33931-1 (ebk)

DOI: 10.1201/9781003339311

Contents

Preface

The steady growth of cellular and wireless communications motivates researchers to improve the performance of the systems, overcome the limitations and face new the challenges. One of the key blocks in a wireless radio is the RF oscillator which its purity limits the radio performance. The oscillator's phase noise in a transmit chain results in power leakage into adjacent channels. In the receive chain, the downconversion of a large interferer with noisy local oscillator (LO) cause reciprocal mixing. Furthermore, in orthogonal frequency-division multiplexing (OFDM) systems, the phase noise leads to inter carrier interference and a degradation in the digital communication bit error rate. The trade-off between oscillator's phase noise and its power consumption introduce a challenge for oscillator designers.

The main focus of this book is on the design and implementation of RF oscillators for wireless (mostly cellular) applications. Each oscillator that is introduced in these chapters tackles an obstacle in RF designs, such as low $1/f^2$ or low $1/f^3$ phase noise requirements, low voltage, low power requirements, and wide tuning range requirements.

Chapter 1 discusses how a transceiver performance can be limited by an oscillator characteristics. It also reviews how technology scaling affects an oscillator's performance.

Chapter 2 is a reminder how circuit noise up-converts to phase noise in an oscillator, and then briefly introduces and compares different LC oscillator structures.

In Chapter 3 we introduce a class-F_3 oscillator topology which demonstrates an improved phase noise performance by enforcing a pseudo-square voltage waveform around the LC tank by increasing the third harmonic of the fundamental oscillation voltage through an additional impedance peak. Furthermore, a comprehensive study of circuit-to-phase-noise conversion mechanisms of different classes of RF oscillator is presented.

In Chapter 4, we elaborate on a design and implementation of class-F_2 oscillators. The main idea is to enforce a clipped voltage waveform around the LC tank by increasing the second-harmonic of fundamental oscillation voltage through an additional impedance peak, thus giving rise to a class-F_2 operation. This oscillator specifically addresses the ultra-low phase noise design space while maintaining high power efficiency. Extensive experimental results are also presented at the end of this chapter.

Excited by a harmonically rich tank current, a typical oscillation voltage waveform is observed to have asymmetric rise and fall times. This results in an effective impulse sensitivity function (ISF) of a non-zero dc value, which facilitates the flicker ($1/f$) noise up-conversion into the oscillator's $1/f^3$ phase noise. Chapter 5 elaborates a method to reduce a $1/f$ noise up-conversion in voltage-biased RF oscillators.

Chapter 6 introduces and analyzes in detail an oscillator with switching current sources to reduce supply voltage and power without sacrificing its phase noise and startup margins. This oscillator is specifically addressed IoT application constraints.

In Chapter 7 a method to broaden a tuning range of an LC-tank oscillator without sacrificing its area is presented. The extra tuning range is achieved by forcing a strongly coupled transformer-based tank into a common-mode resonance at a much higher frequency than in its main differential-mode oscillation. The oscillator employs separate active circuits to excite each mode but it shares the same tank, which largely dominates the core area but is on par with similar single-core designs.

Chapter 8 presents a design guide to estimate the time dependent dielectric beak down of any analog circuit with evaluating life time of class-F oscillators as an example.

List of Figures

List of Tables

List of Abbreviations

4G	Fourth generation
5G	Fifth generation
AM	Amplitude modulation
BEOL	Back-end-of-line
BTS	Base station
BLE	Bluetooth Low Energy
CMP	Chemical-mechanical polishing
CMOS	Complementary metal-oxide-semiconductor
DCO	Digitally controlled oscillator
DT	Direct quantum-mechanical tunneling
FoM	Figure of merit
FinFET	Fin Field-effect transistor
FN	Fowler–Nordheim
GSM	Global system for mobile
ISF	Impulse sensitivity function
IoT	Internet-of-Things
KCL	Kirchhoff's current law
LTV	Linear time variant
LO	Local oscillator
MoM	Metal-oxide-metal
MOS	Metal-oxide-semiconductor
NBTI	Negative bias temperature instability
OFDM	Orthogonal Frequency-Division Multiplexing
PER	Packet error rate
PLL	Phase lock loop
PM	Phase modulation
PN	Phase noise
PA	Power amplifier
PVT	Process-voltage-temperature
Q-factor	Quality factor
RF	Radio frequency

SNR	Signal-to-noise ratio
SoC	System-on-chips
TDDB	Time-dependent dielectric breakdown
TR	Tuning range
ULP	Ultra-low power
UMTS	Universal mobile telecommunication system
VCO	Voltage control oscillator

1

Introduction

1.1 Introduction

While mobile phones enjoy the largest production volume ever of any consumer electronics products, the demands they place on RF/mm-wave transceivers are particularly aggressive, especially on integration with digital processors, low area, low power consumption, while being robust against process-voltage-temperature (PVT) variations. Figure 1.1 (a) illustrates the evolution of data rates for wireless LAN, cellular, and wireline short links over time [1]. Interestingly, there is a constant ~10× increase in bit rate every five years for both wireline and wireless links. Since mobile terminals inherently operate on batteries, their power budget is basically constant. Hence, an ever-decreasing power per bit is required to maintain the system lifetime. As shown in Figure 1.1 (b), the RF front-end circuitry consumes significant power for typical use cases of mobile terminals such as voice call, web browsing and email [2]. Consequently, power efficiency of RF building blocks has become a major issue, especially when designing system-on-chips (SoC) for wireless communications.

On the other hand, in the upcoming years, the wireless Internet-of-Things (IoT) will enable more objects to be sensed and controlled remotely, realizing more integration between the physical and digital worlds. According to communication giant Cisco systems [3] there will be approximately 50 billion Internet-connected objects (things) by 2020, just 2.7 percent of all the things that will be on the planet. However, the system lifetime still tends to be severely limited by its power consumption, available battery technology and efficiency of its energy harvester. Consequently, the key challenge of these wireless sensors is the ability to function at the lowest power possible while being robust to PVT variations. Similarly, the power consumption of RF building blocks should be reduced to satisfy the lifetime demands of IoT systems. Furthermore, RF circuitries such as oscillator and phase lock loop

1

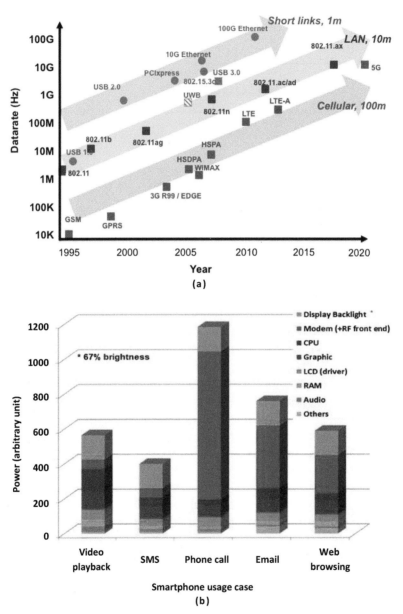

Figure 1.1 (a) Evolution of data rates for wireless LAN, cellular, and wireline short links over time [1]; (b) power usage in a smartphone [2].

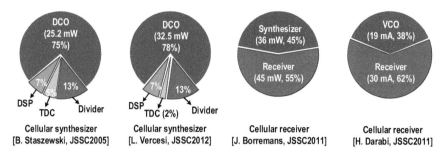

Figure 1.2 Contribution of RF oscillator to the power consumption of cellular frequency synthesizers and receivers.

(PLL) should be able to turn on/off quickly to permit high energy-efficiency during intermittent operation.

The RF oscillator is the second most power hungry block of a wireless radio (after power amplifiers). As shown in Figure 1.2, the RF oscillator consumes a disproportionate amounts of the power of a cellular frequency synthesizer [4, 5] and burns more than 30% of the cellular receiver power [6, 7]. Consequently, any power reduction in an RF oscillator will greatly benefit the overall power efficiency of the cellular transceiver. For IoT applications, the commercial perspective is now focusing on Bluetooth Low Energy (BLE). In the state-of-the-art BLE radios [8–10], the PLL power consumption is merely 3–4× lower than that of power amplifier (PA) at the maximum BLE output power of 1 mW. However, the frequency synthesizer activity is much longer than that of a PA, making the PLL the most energy-hungry block in a BLE transceiver. Consequently, RF oscillators, as one of the BLE transceiver's most power hungry circuitry, must be very power efficient.

On the other hand the RF oscillators' purity limits the transceiver performance. The oscillator's phase noise in a transmit chain results in power leakage into adjacent channels. In the receive chain, the down-conversion of a large interferer with a noisy local oscillator (LO) causes reciprocal mixing. Furthermore, in orthogonal frequency-division multiplexing (OFDM) systems, the phase noise leads to inter-carrier interference and a degradation in the digital communication bit error rate. Table 1.1 summarizes the frequency bands and the phase noise requirement specifications for some communication standards. The trade-offs between oscillator's phase noise and its power consumption introduce a challenge for oscillator designers. To achieve high frequency accuracy, oscillators are incorporated in a PLL (as is shown in Figure 1.3 for both analog and digital PLLs), they can benefit from high pass nature of filtering of their noise by the loop (see Figure 1.4). This reduction

Table 1.1 Communication standards requirements [11]

Standard	Frequency Band (GHZ)	Required Phase Noise (dBc/HZ)
Bluetooth	2.402–2.480	–84 @ 1 MHz –114 @ 2 MHz –129 @ 3 MHz
GSM 900/1800	0.880–0.960 1.710–1.880	−122 @ 0.6 MHz −132 @ 1.6 MHz −139 @ 3 MHz
UMTS	1.920–2.170 1.900–2025	−132 @ 3 MHz −132 @ 10 MHz −144 @ 15 MHz
WiFi	2.412–2.472 5.150–5.350 5.470–5.825	−102 @ 1 MHz −125 @ 25 MHz

of the oscillator's low frequency noise in the synthesizer is highly dependent on the loop bandwidth. The loop bandwidth of the PLL is usually chosen to minimize the noise contribution of the frequency reference and charge pumps. However, if this bandwidth is less than the $1/f^3$ corner of the oscillator then part of the oscillator's low frequency noise remains unfiltered.

Another challenge for the recent RF oscillator designers is the ability to design a wide tuning range oscillator while having low phase noise. The multi-standard applications that are now trending demand such oscillators. The trade-off between the quality factor of the switch capacitor bank that is tuning the LC oscillators and the oscillator's tuning range is the obstacle in wide tuning-range oscillator design. The MOS transistor switch introduces a resistance that defines the switched capacitor bank's quality factor in on-state, consequently a lower resistance and so a larger MOS transistor is required for phase noise consideration. However, in the off-state, the series combination of the capacitor in the tank and the switch's parasitic capacitances defines the equivalent tank capacitance. Consequently a smaller switch is preferred to increase the tuning range. This trade-off makes it impossible to meet both wide tuning range and low phase noise at the same time. For a moderate phase noise, the tuning range of the oscillator can hardly go beyond 50% [13]. Some designers tried to switch inductors or transformers instead of the capacitors in order to increase tuning range, however the equivalent tank's Q-factor and consequently phase noise is degraded due to the switches in the signal path. Furthermore, due to the reduced oscillation voltage that is tolerable by the nano-metric oxide thickness of advanced technologies CMOS process, the low phase noise design is even more challenging.

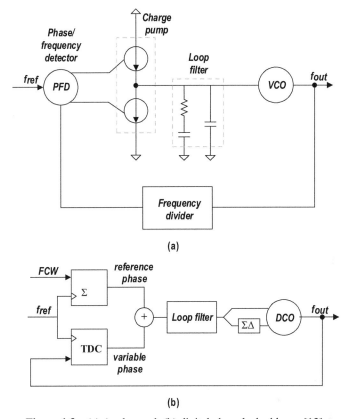

Figure 1.3 (a) Analog and; (b) digital phase locked loops [12].

In this book, the main goal is to to elaborate implementation of innovative RF oscillator structures that demonstrate better PN performance, lower cost, and higher power efficiency than the traditional architectures do.

1.2 Technology Scaling

The size, cost, and power consumption of digital circuits are reduced by technology scaling. However, the design of analog and RF circuits faces many difficulties using more advanced CMOS technologies. Consequently, in the semiconductor industry, there are two divergent trends for choosing a technology node for fabricating analog circuits. One trend is to implement analog circuits in an older technology to exploit a higher voltage headroom. This approach is chiefly used in medical, automotive, and high-efficiency

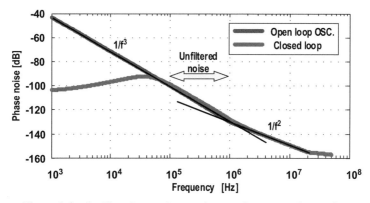

Figure 1.4 Oscillator's open loop and output frequency phase noise.

lighting applications [14]. Another trend is to implement analog and digital circuits together in the most advanced node such as a 16-nm FinFET. The approach is dictated by the market to achieve highest digital performance with lowest fabrication cost.

The vision for both wireless cellular communication and the Internet-of-Things keeps moving towards the second strategy [15]. For example, a few years ego IMEC published the first integrated wireless sensor node in 40-nm CMOS including a microcontroller, digital baseband, power management, and BLE transceiver [8]. Furthermore, Intel and DMCE presented a SAW-less HSPA transceiver with on-chip integrated 3G power amplifiers in 65-nm at ISSCC 2015 to enable real low-cost monolithic system integration [16]. Consequently, it is instructive to investigate the effects of technology scaling on the performance of an RF/mm-wave oscillators.

1.2.1 Supply Voltage

To continue implementing increasingly complex functions while reducing the overall solution costs, scaling of CMOS transistors is inevitable. As circuits become denser, all of the physical dimensions of the transistors must be reduced correspondingly. The SiO_2 oxide-layer thickness reduction is accompanied by migrating to smaller supply voltages (see Figure 1.5). This is to maintain the electric field strength across the oxide in order to prevent the device performance degradation due to the time dependent dielectric breakdown (TDDB) [17]. The supply voltage, V_{DD}, is reduced while RF and analog circuits must maintain their dynamic range, noise performance and output power. For example, the oscillator phase noise performance degrades

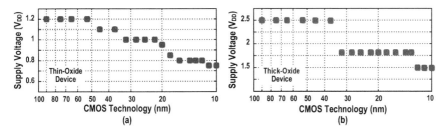

Figure 1.5 Nominal supply voltage versus CMOS technology node for (a) thin-oxide and (b) thick-oxide devices.

by 6 dB/octave with the reduction of their supply voltage [18]. To compensate this phase noise penalty, the equivalent input parallel resistance of an LC tank should be proportionally decreased by reducing the tank's inductance. However, the resistance of the tank's interconnects will start dominating the resonator losses and, consequently, the effective Q-factor and oscillator power efficiency will dramatically drop.

1.2.2 Quality Factor of Passives

Most RF CMOS oscillators employ passive components such as integrated inductors, transformers, and capacitors to realize on-chip LC tanks. Generally, top thick metal layers are used for the realization of inductive components while thinner lower-level metals are exploited in metal-oxide-metal (MoM) capacitors. Note that the $0.18\,\mu$m node was the last generation of 8-inch wafer processes with aluminum metallization as almost all modern processes use copper metallization that improves the resistive loss and current handling capability of passive components. Consequently, the quality factor of passives was improved when RF products migrated from $0.18\,\mu$m to $0.13\,\mu$m technology.

By migrating to more advanced nano-scale CMOS technologies, however, the thickness of lower level metals and interlayer dielectric layers reduce correspondingly as shown in Figure 1.6 (a). As a consequence, the physical dimension of a given capacitance becomes smaller but with a worse quality factor (see Figure 1.6 (b)). Fortunately, the thickness of top thick metal layers almost remains constant with scaling. However, the top-metal is closer to the lossy substrate. Hence, the capacitive parasitic, self-resonant frequency, and quality factor of inductor/transformer slightly degrade with scaling as shown in Figure 1.6 (c).

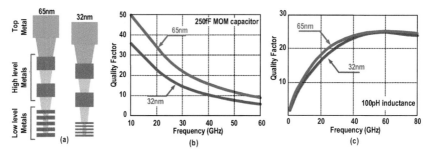

Figure 1.6 (a) Back-end-of-line (BEOL) metallization; quality factor of (b) a 250 fF capacitor, and (c) a 100 pH inductor in 65 nm and 32 nm CMOS technologies [19].

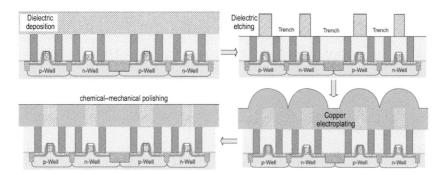

Figure 1.7   Damascene process steps [20].

As mentioned above, in most advanced CMOS technologies, copper is used for interconnections due to its low sheet resistance, high maximum current density, high thermal conductivity and resilience to electromigration failures. However, it is difficult to pattern copper using conventional etching techniques. Consequently, unlike traditional metallization of aluminum, copper metallization needs an additional damascene process as shown in Figure 1.7 [20]. The inter-level dielectric is first deposited in the damascene process. Secondly, the dielectric is etched to define trenches where the metal lines will lie. Thirdly, copper is electroplated to fill the patterned oxide trenches. Finally, the surface is planarized and polished to remove surplus copper outside the desired metal lines using chemical-mechanical polishing (CMP). Unfortunately, CMP suffers from dishing and erosion phenomena. Since copper is much softer than the inter-level dielectric, areas with higher metal density are polished much faster than the others. Consequently, the metal thickness of the sparse areas become thicker than that of the dense

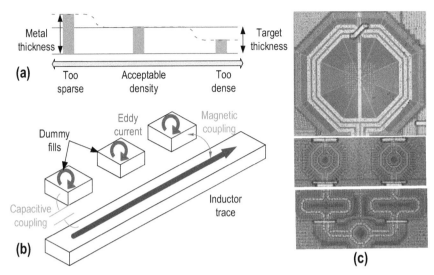

Figure 1.8 (a) Thickness variation by erosion in the CMP stage; (b) electromagnetic coupling between the wire and dummy fills; (c) inductor/transformer with lots of dummy metal fills.

places as shown in Figure 1.8 (a) [22]. To resolve this issue, a minimum metal density must be satisfied for the entire chip. For example, the minimum metal density must be at least 25% in any $100\,\mu\text{m}\times100\,\mu\text{m}$ square in a 28-nm technology. Hence, inductors and transformers must include dummy metal pieces from the lowest to the highest metal layer (see Figure 1.8 (c)). Metal dummies show negligible effects on the windings self-inductance and the coupling factor. However, as shown in Figure 1.8 (b), eddy currents in dummy fills increase the loss and thus the inductor's/transformer's Q-factor could be degraded by 10% [23].

To conclude this section, the quality factor of passives degrades by migrating to more advanced nano-scale CMOS technologies. It leads to a worse phase noise performance for oscillators.

1.2.3 Noise of Active Devices

Figure 1.9 (a) shows the normalized input-referred 1/f noise, S_{vg}, versus frequency for different technology nodes. Due to oxide scaling enabled by high-k/metal gate technologies, the normalized S_{vg} is monotonically reduced with each successive technology node. For the same transistor area, S_{vg}

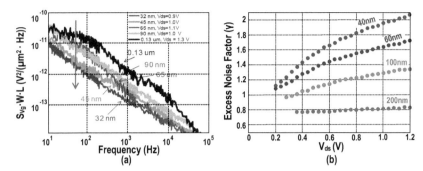

Figure 1.9 (a) Flicker noise scaling trend; (b) measured excess noise (γ) factor versus drain–source voltage at 10 GHz and $V_{gs} = 1.0$ V for different gate lengths of NMOS transistors in 40 nm LP technology [21].

decreases $\sim 10\times$ from the 0.13 μm node to 32 nm node. However, as shown in Figure 1.9 (b), the excess noise factor of CMOS transistors increases by migrating to finer CMOS technologies. Consequently, oscillator's core transistors inject more thermal noise to the tank, degrading its phase noise performance.

References

[1] S. Kosonocky, "Indicators – Historic trends in technical themes digital systems" *IEEE International Solid-State Circuits Conference (ISSCC) Trends*, Feb. 2015, pp. 1–46.

[2] G. Yeap, "Embracing the internet of everything To capture your share of $14.4 trillion" *IEEE International Electron Devices Meeting (IEDM)*, 1.3.1–1.3.8, pp. 541–544.

[3] J. Bradley, J. Barbier, and D. Handler, "Indicators – Historic trends in technical themes digital systems" *Cisco Systems Inc.*, 2013, pp. 1–18. Available: http://www.cisco.com/web/about/ac79/docs/innov/IoE_Economy.pdf.

[4] R. B. Staszewski, J. L. Wallberg, S. Rezeq, C.-M. Hung, O. E. Eliezer, S. K. Vemulapalli, C. Fernando, K. Maggio, R. Staszewski, N. Barton, M.-C. Lee, P. Cruise, M. Entezari, K. Muhammad, and D. Leipold, "All-digital PLL and transmitter for mobile phones," *IEEE J. Solid-State Circuits*, vol. 40, no. 12, pp. 2469–2482, Dec. 2005.

[5] L. Vercesi, L. Fanori, F. D. Bernardinis, A. Liscidini, and R. Castello,' "A dither-less all digital PLL for cellular transmitters," *IEEE J. Solid-State Circuits*, vol. 47, no. 8, pp. 1908–1920, Aug. 2012.

[6] J. Borremans, G. Mandal, V. Giannini, B. Debaillie, M. Ingels, T. Sano, B. Verbruggen, and J. Craninckx, "A 40nm CMOS 0.4-6 GHz receiver resilient to out-of-band blockers," *IEEE J. Solid-State Circuits*, vol. 46, no. 7, pp. 1659–1671, Jul. 2011.

[7] H. Darabi, P. Chang, H. Jensen, A. Zolfaghari, P. Lettieri, J. C. Leete, B. Mohammadi, J. Chiu, Q. Li, S.-L. Chen, Z. Zhou, M. Vadipour, C. Chen, Y. Chang, A. Mirzaei, A. Yazdi, M. Nariman, A. Hadji-Abdolhamid, E. Chang, B. Zhao, K. Juan, P. Suri, C. Guan, L. Serrano, J. Leung, J. Shin, J. Kim, H. Tran, P. Kilcoyne, H. Vinh, E. Raith, M. Koscal, A. Hukkoo, V. R. C. Hayek, C. Wilcoxson, M. Rofougaran, and A. Rofougaran, "A quad-band GSM/GPRS/EDGE SoC in 65nm CMOS," *IEEE J. Solid-State Circuits*, vol. 46, no. 4, pp. 872–882, Apr. 2011.

[8] Y.-H. L., C. Bachmann, X. Wang, Y. Zhang, A. Ba, B. Busze, M. Ding, P. Harpe, G.-J. van Schaik, G. Selimis, H. Giesen, J. Gloudemans, A. Sbai, L. Huang, H. Kato, G. Dolmans, K. Philips, and H. de Groot, "A 3.7mW-RX 4.4mW-TX fully integrated Bluetooth Low-Energy/IEEE802.15.4/proprietary SoC with an ADPLL-based fast frequency offset compensation in 40nm CMOS," *IEEE International Solid-State Circuits Conference Digest of Technical Papers (ISSCC)*, Feb. 2015, pp. 236–237.

[9] J. Prummel, M. Papamichail, J. Willms, R. Todi, W. Aartsen, W. Kruiskamp, J. Haanstra, E. Opbroek, S. Rievers, P. Seesink, J. van Gorsel, and H. Woering, "A 10 mW Bluetooth Low-Energy Transceiver With On-Chip Matching," *IEEE J. Solid-State Circuits*, vol. 50, no. 12, pp. 30773088, Dec. 2015.

[10] T. Sano, M. Mizokami, H. Matsui, K. Ueda, K. Shibata, K. Toyota, T. Saitou, H. Sato, K. Yahagi, and Y. Hayash, "A 6.3 mW BLE transceiver embedded RX image-rejection filter and TX harmonic-suppression filter reusing on-chip matching network," *IEEE International Solid-State Circuits Conference Digest of Technical Papers (ISSCC)*, Feb. 2015, pp. 240–241.

[11] F. Pepe, "Analysis and minimization of flicker noise up-conversion in radio-frequency LC-tuned oscillators" PhD dissertation.

[12] R. B. Staszewski, "State-of-the-art and future directions of high-performance all-digital frequency synthesis in nanometer CMOS," *IEEE Trans. Circuits Syst. I, Reg. Papers*, vol. 58, no. 7, pp. 1497–1510, Jul. 2011.

[13] G. Li, L. Liu, Y. Tang, and E. Afshari, "A low phase-noise wide tuning-range oscillator based on resonant mode switching," *IEEE J. Solid-State Circuits*, vol. 47, no. 6, pp.1295–1308, June 2012.

[14] A. Parssinen, "Indicators – Historic trends in technical themes analog systems" *IEEE International Solid-State Circuits Conference (ISSCC) Trends*, Feb. 2015, pp. 1–46.

[15] A. Thomsen, "Indicators – Historic trends in technical themes communication systems" *IEEE International Solid-State Circuits Conference (ISSCC) Trends*, Feb. 2015, pp. 1–46.

[16] J. Moreira et al., "single-chip HSPA transceiver with fully integrated 3G CMOS power amplifiers," *IEEE International Solid-State Circuits Conference Digest of Technical Papers (ISSCC)*, Feb. 2015, pp. 162–163.

[17] E. Wu, E. Nowak, W. Vayshenker, A. Lai, and D. Harmon, "CMOS scaling beyond the 100-nm node with silicon-dioxide-based gate dielectrics," *IBM Journal of Research and Development*, vol. 46, no. 2/3, pp. 287–1308, Mar/May 2002.

[18] B. Razavi, *RF Microelectronics.* Prentice Hall PTR, 2011. Available: http://books.google.nl/books?id=_TccKQEACAAJ&hl.

[19] E. Mammei, E. Monaco, A. Mazzanti, and F. Svelto, "VCO with 177.5dBc/Hz minimum noise FOM using inductor splitting for tuning extension," *IEEE International Solid-State Circuits Conference Digest of Technical Papers (ISSCC)*, Feb. 2013, pp. 350–351.

[20] R. F. Yanda, M. Heynes, and A. K. Miller, *Demystifying Chipmaking.* Elsevier, 2005. [Online]. Available: http://store.elsevier.com/ Demystifying-Chipmaking/ Richard-Yanda/isbn-9780080477091/.

[21] G. Smit, A. Scholten, R. Pijper, L. Tiemeijer, R. van der Toorn, and D. Klaassen, "RF-Noise Modeling in Advanced CMOS Technologies," *IEEE Transactions on Electron Devices*, vol. 61, no. 2, pp. 245–354, Feb. 2014.

[22] A. Tsuchiya and H. Onodera, "Patterned floating dummy fill for on-chip spiral inductor considering the effect of dummy fill," *IEEE Transactions on Microwave Theory and Techniques*, vol. 56, no. 12, pp. 3217–3222, Dec. 2008.

[23] F.-W. Kuo, R. Chen, K. Yen, H.-Y. Liao, C.-P. Jou, F.-L. Hsueh, M. Babaie, and R. B. Staszewski, "A 12 mW all-digital PLL based on class-F DCO for 4G phones in 28 nm CMOS", *Proceedings of IEEE VLSI Circuits Symposium*, 2014, pages 1–2.

2

LC Oscillator Structures

2.1 Introduction

The oscillators are the only block that is universally used in both transmit and receive paths (see Figure 2.1), and consequently their spectral purity and efficiency highly affect the transceiver performance. The phase noise of the oscillator results in reciprocal mixing in the receive path, where the blocker is mixed with the oscillator's phase noise and shows itself on top of the desired signal and consequently degrades the receiver sensitivity [1]. This problem especially shows itself in contemporary mobile phones that support 2G, 3G, and 4G modes and WiFi standards with two very close antennas in one device [2], or in wide-band CMOS receivers without off-chip SAW filters, in which blockers can enter the chip without any pre-attenuation [3]. In transmit path, the amplified phase noise of the transmitter's oscillator can desensitize a nearby receiver [1]. Furthermore, as one of the most power hungry blocks in the transceiver, its power consumption limits the efficiency of the transceiver [4, 5]. Therefore, understanding and modeling the phase noise of an oscillator have been the subject of numerous studies [6–12]. The linear time-variant model through the impulse response of each noise source of the oscillator [10] is the most approached method since its introduction.

We are relying on this method to analyze the oscillators in this book; so let us have a quick overview first. The relatively accurate modeling of phase noise in this method is by acknowledging time-variant behavior of the oscillators. To make it more clear, note that a current impulse injected to the tank of Figure 2.2(a) can change oscillation phase and/or amplitude depending on the injection time (see Figure 2.2(b,c)). If the current impulse is injected when oscillation waveform is at its maximum, the oscillation amplitude will be disturbed but the phase will not be. On the other hand, current impulse at the zero-crossings results in a minimum amplitude and maximum phase disturbance. The impulse response is however periodic with respect

13

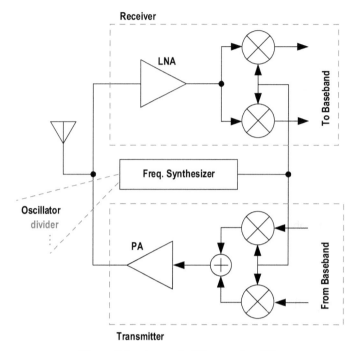

Figure 2.1   A generic RF transceiver [1].

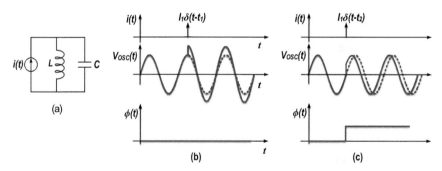

Figure 2.2   Phase response to an impulse current [10].

to injection time. The impulse sensitivity function (ISF), $\Gamma(\omega_0\tau)$, is defined as a dimensionless, periodic function with period of 2π that describes the oscillation phase shift from injected current impulses during the period [10]. ISF is a periodic function and, consequently, can be written in a Fourier series,

$$\Gamma(\omega_0\tau) = \frac{c_0}{2} + \sum_{i=1}^{\infty} c_n \cos(n\omega_0 t + \theta_n). \tag{2.1}$$

Phase modulation is then obtained by convolving the current noise source and ISF as

$$\phi_n(t) = \frac{1}{q_{max}} \left[\frac{c_0}{2} \int_{-\infty}^{t} i(\tau) + \sum_{i=1}^{\infty} c_n \int_{-\infty}^{t} i(\tau) \cos(n\omega_0 t + \theta_n) \right], \quad (2.2)$$

where q_{max} is the maximum charge displacement at the capacitance of the node that the noise is injected.

For a current such as $i(t) = I_n \cos[(n\omega_0 + \Delta\omega)t]$, the excess phase can be found as

$$\phi(t) \approx \frac{I_n c_n \sin(\Delta\omega)}{2q_{max}\Delta\omega}. \quad (2.3)$$

The modulated phase shows itself in the phase noise spectrum since we can write

$$x(t) = A\cos(\omega_0 t + \phi_n(t)) \approx A\cos(\omega_0(t)) - A\phi_n(t)\sin(\omega_0 t), \quad (2.4)$$

and consequently this injected current results in two sidebands at $\omega_0 \pm \Delta\omega_0$ and

$$\mathcal{L}(\Delta\omega) = 10\log_{10}\left(\frac{I_n c_n}{4q_{max}\Delta\omega}\right)^2. \quad (2.5)$$

The same method can be generalized for random noise sources and by applying the Parseval's relation to derive the phase noise for a white power spectral density noise as,

$$\mathcal{L}(\Delta\omega) = 10\log_{10}\left(\frac{\frac{\overline{i_n^2}}{\Delta f}\frac{1}{2\pi}\int_0^{2\pi}\Gamma^2(\phi)d\phi}{4q_{max}^2\Delta\omega^2}\right). \quad (2.6)$$

The most accurate method to calculate ISF of each noise source is by simulation. An impulse current should be injected to a node in the circuit at a certain time. The time shift of the oscillation should be measured after a few cycles and be converted to the phase shift. By sweeping the injection time of the current impulse over one oscillation period, ISF can be measured. Very recently, a fast and accurate simulation technique of ISF based on positive sidebands of periodic transfer function (e.g. PXF in Cadence) was revealed in [13].

Upconversion of the device's 1/f noise to phase noise can also be investigated by this method.

If the application demands a low phase noise, the LC oscillator structure should be chosen. The thermal to phase noise upconversion (20 dB/dec region) of these oscillators can be found as

$$
\begin{aligned}
\mathcal{L}(\Delta\omega) &= 10\log_{10}\left(\frac{R_t kT}{2Q_t^2 V_{OSC}^2} \cdot F \cdot \left(\frac{\omega_0}{\Delta\omega}\right)^2\right) \\
&= 10\log_{10}\left(\frac{kT}{2\,Q_t^2\,\alpha_I\,\alpha_V\,P_{DC}} \cdot F \cdot \left(\frac{\omega_0}{\Delta\omega}\right)^2\right),
\end{aligned}
\tag{2.7}
$$

where R_t is the equivalent parallel resistance of the tank, k is Boltzmann's constant, T is the temperature, $\alpha_V = \frac{V_{osc}}{V_{DD}}$ and $\alpha_I = \frac{I_{\omega_0}}{I_{DC}}$, and F is the noise factor and can be found as

$$
F = \sum_i \frac{R_t}{2kT} \cdot \frac{1}{2\pi} \int_0^{2\pi} \Gamma_i^2\,(\phi)\,\overline{i_{n,i}^2(\phi)}\,d\phi,
\tag{2.8}
$$

in which Γ_i is the ISF of the ith noise source.

2.2 Class-B Oscillator Topology

The traditional class-B oscillator, of Figure 2.3(a), is widely used in RF applications due to its simplicity and robustness. The noise factor in a class-B structure is ideally equal to $\gamma + 1$ [12] if M_T tail current transistor is an ideal current source. In this case, not only the current source avoids contributing to phase noise, but it also provides an infinite impedance at the common source of the g_m transistors, which, as we explain later, is beneficial for the phase noise reduction. Let us investigate how the performance of this oscillator topology can be improved. The figure of merit (FoM) that is widely used to compare the oscillator performance is

$$
FoM = |PN| + 20\log_{10}(\omega_0/\Delta\omega) - 10\log_{10}(P_{DC}/1mW).
\tag{2.9}
$$

The objective is to reduce the phase noise and/or power consumption of the oscillator.

Increasing the tank's quality factor reduces the oscillator's phase noise. The tank's quality factor depends on both the inductive and capacitive quality factors:

$$
\frac{1}{Q_t} = \frac{1}{Q_L} + \frac{1}{Q_C}.
\tag{2.10}
$$

The inductor's quality factor, Q_L, which usually limits Q_t, is mostly technology-dependent but does not improve with technology scaling. The

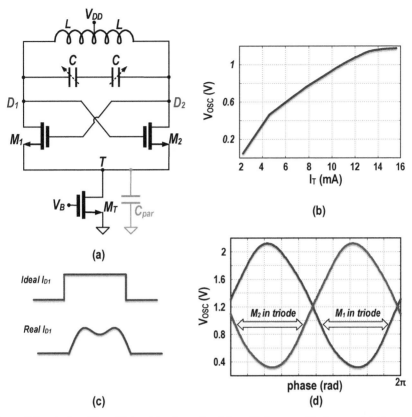

Figure 2.3 A class-B oscillator (a) schematic; (b) oscillation amplitude versus tail current; (c) ideal and real drain current waveforms; (d) oscillation voltage waveforms.

capacitive quality factor, Q_C, on the other hand, depends on the tuning range of the oscillator. The switched-capacitor structure shown in Figure 2.4 is often used to tune the conventional oscillators. When M_s is on, $C_{on} = \frac{C}{2}$, and the switch's on-resistance, r_{on} defines $Q_C = \frac{1}{2r_{on}C\omega}$. To improve Q_c, r_{on} and consequently M_s size should increase. However, a larger M_s would add to the parasitic capacitance and consequently would increase the switched-capacitor equivalent capacitance when M_s is off: $C_{off} = \frac{CC_{par}}{2(C+C_{par})}$. Consequently, Q_t will be defined by the technology and oscillator's tuning range, and is rarely a design parameter to substantially improve the phase noise.

Another approach to improve the oscillator's phase noise is by reducing the tank's inductance while keeping its quality factor the same. Doing so

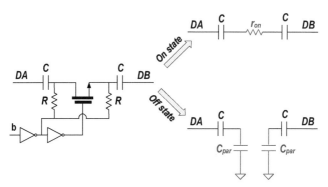

Figure 2.4   The switch capacitor tuning circuit in on and off states.

reduces $R_t = L\omega Q_t$; however, it increases the power consumption $P_{DC} = \frac{V_{OSC}^2}{\alpha_V \alpha_I R_t}$, with the same rate, and hence FoM will not improve. Furthermore, by reducing the inductor's size, the tank interconnection losses become more important and, at some point, they will limit its quality factor. This oscillator shows the best performance when its oscillation amplitude is around V_{DD} [14–16] and consequently $\alpha_V = 1$. After this point, the oscillation amplitude stops increasing with the tail current increase (see Figure 2.3(b)) while its power consumption still increases linearly with the tail current, thus reducing the FoM. The drain current of $M_{1,2}$ transistors has almost a square waveform when the current source is ideal and so $\alpha_I = \frac{2}{\pi}$ (see Figure 2.3(c)). However, in a real scenario, a non-ideal current source will bring up some issues and limitations. First of all, M_T transistor will contribute to the phase noise, thus increasing the noise factor beyond $1 + \gamma$. The minimum tail node voltage, V_T, is also limited by the need to keep the M_T transistor in saturation; consequently the maximum oscillation voltage amplitude reduces to $V_{DD} - V_{sat}$ and so $\alpha_V < 1$ ($\alpha_V \approx 0.8$). The capacitance at node T tends to keep its voltage at a constant level. Consequently, for large oscillation amplitudes with $M_{1,2}$ entering the triode region, the ideal square wave of $M_{1,2}$ drain current experiences a dimple, as shown in Figure 2.3(c). Hence, α_I drops from the ideal value of $\frac{2}{\pi}$, and phase noise is increased. On the other hand, when M_1 or M_2 transistors enter triode region for a portion of oscillation period, they will show a low impedance. Furthermore, the equivalent parasitic capacitance at node T creates a low impedance path to ground. Therefore, the tank finds a discharge path to the ground for the time that either one of these transistors is in the triode region and so its quality factor drops, thus limiting the oscillator's phase noise. The M_T transistor size is usually relatively large in order to reduce its flicker noise. Consequently, the parasitic capacitor at node T is large enough to provide such a low frequency path. However, it is also helpful in partially filtering the M_T transistor's thermal noise.

Various solutions are proposed in the literature to improve the class-B topology phase noise or to improve the oscillator's phase noise–power consumption trade-off by introducing new classes of oscillations. One of the most effective techniques that could improve the class-B considerably is the noise filtering technique [17]. In this technique, the M_T's thermal noise is filtered by a relatively large capacitor and a high impedance path is inserted between the core transistors and M_T to deny any discharge path to the tank. Although this technique is very effective, since the high impedance path is realized by another resonator, it increases the die area significantly. Another interesting technique to improve the oscillator's phase noise is to couple N oscillator cores together [18]. This technique has been used in microwave circuits [19] and also employed to improve phase noise in RF applications [20]. With coupling N cores, phase noise reduces by a factor of N while the power consumption increases by the same factor. Consequently, although the phase noise is reduced, the FoM remains the same. However, the die area is surely getting N times larger.

In the following sections, we briefly review other oscillator topologies that attempt to improve the phase noise–power consumption trade-off in an oscillator. In a class-C structure, $M_{1,2}$ are biased in a way as to always remain in saturation during the whole oscillation period. In another strategy, the oscillation waveforms in class-D and class-F structures offer special impulse sensitivity functions (ISFs) that prevent circuit noise from upconverting to phase noise.

2.3 Class-C Oscillator Topology

Class-C structure [21] is shown in Figure 2.5(a). In this class of operation, the core transistors are kept in saturation and, consequently, they show a high impedance during the entire oscillation period. The tank does not find a discharge path to the ground and so its quality factor is preserved. This structure also saves 36% of the power consumption for the same phase noise by changing the square pulses of the $M_{1,2}$ drain current in class-B operation to narrow and tall pulses with $\alpha_I = 1$. To ensure the saturation region of operation, $M_{1,2}$ transistor gates are decoupled from the tank's oscillation voltage and are biased at a value well below the V_{DD} voltage. A large

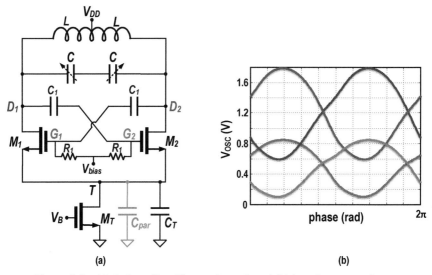

Figure 2.5   (a) A class-C oscillator schematic and (b) its voltage waveforms.

capacitor in parallel with the M_T current source allows class-C like tall and narrow current pulses for the $M_{1,2}$ transistors.

However, the maximum oscillation amplitude is limited in this topology. If the oscillation amplitude gets large enough to push $M_{1,2}$ into triode region, not only the tank's quality factor heavily drops due to large C_T, but also $M_{1,2}$ drain current will no longer be tall and narrow pulses and α_I dramatically drops. Consequently, although the phase noise and power efficiency are improved for low oscillation amplitudes compared to class-B oscillator structure with the same amplitude, the minimum achievable phase noise of this structure is limited. An attempt to increase the class-C swing is done by removing the current source transistor M_T and generating V_{bias} by a current mirror circuit [22]. That oscillator topology also suffers from a trade-off between its robust start-up and the maximum oscillation voltage in steady state [23]. V_{bias} should be relatively large to facilitate start-up, but large V_{bias} values limit the steady-state oscillation amplitude. It was proposed to adjust V_{bias} dynamically in a negative feedback loop [23–25], which consumes extra power (see Figure 2.6(a)), or to employ class-B switching transistors in parallel with the class-C ones to ensure start-up for low V_{bias} values [26, 27], which reduces α_I and consequently power efficiency (see Figure 2.6(b)). The power efficiency of this structure motivated designers in [28] to incorporate this oscillator topology in a Bluetooth low energy (BLE) transmitter.

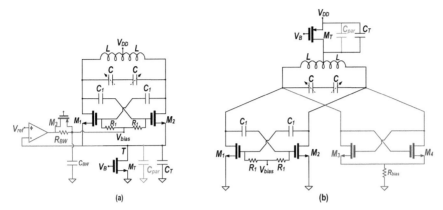

Figure 2.6 (a) A class-C with dynamic generation of V_{bias} [23]; (b) a hybrid class-B/class-C oscillator [27].

Interestingly, the inherently low flicker noise operation of class-C has long eluded proper explanation. It was only very recently explained in [13] by applying the principles disclosed in this book.

2.4 Class-D Oscillator Topology

The schematic of this oscillator topology is shown in Figure 2.7(a). The tail transistor is removed in this structure, eliminating the overhead voltage necessary for the tail current source transistor. Furthermore, $M_{1,2}$ transistor sizes are chosen large enough to become almost ideal switches. The oscillation voltage amplitude is maximized in this structure and reaches about $3V_{DD}$. By doing so, it pushes $M_{1,2}$ transistors deep into the triode region (even more than in the class-B structure) and, consequently, they generate considerable amount of noise. However, as demonstrated in Figure 2.7(b), the oscillation voltages, V_1 or V_2, are forced to ground for almost half the period. V_1 (V_2) is mostly grounded when M_1 (M_2) is in the triode region, and, consequently, the ISF of node D_1 (D_2) is almost zero for most of this period, preventing M_1 (M_2) noise to be upconverted to phase noise.

The idea of voltage-switching oscillators was first proposed in 1959 [29] with a discrete BJT implementation, consequently not suitable for RF applications. However, recent CMOS technologies make excellent switches with reasonable sizes available and consequently this structure is attracting an increasing interest [30–32]. The high oscillation amplitude in this structure makes it suitable for low-voltage low-phase-noise applications [32, 33].

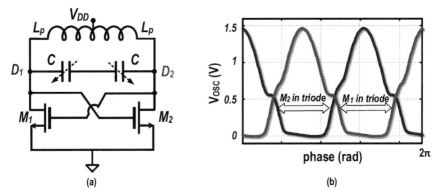

Figure 2.7   (a) A class-D oscillator schematic and (b) its voltage waveforms.

The product of the drain current through MOS switches and voltage is almost zero across the oscillation period, consequently the power efficiency of this structure is beyond 90% [31]. However, this oscillator structure not only *can* work with low voltage supplies but it *should* utilize low power supply voltages, otherwise the $M_{1,2}$ transistors, which should be thin-oxide devices to guarantee nearly ideal switching, will face breakdown. Another limitation of the class-D structure is its relatively high upconversion of low-frequency noise as well as high supply pushing. It has been tried to minimize this problem by an on-chip LDO in [34], but it is power consuming. We elaborate this problem in detail in Chapter 5 and then disclose a solution.

2.5 Class-F Oscillator Topologies

If the ISF of a certain oscillation waveform is negligible for some amount of oscillation period, the circuit noise cannot be upconverted to phase noise during that time, which is beneficial in reducing the oscillator's phase noise. Class-F oscillators realize such oscillation waveforms by giving rise to either *third* or *second* harmonic of oscillation voltage [35–39]. This class of oscillators is discussed in detail in Chapters 3 and 4.

2.6 Conclusion

In this chapter, we briefly introduced different oscillator structures and mentioned their benefits and drawbacks. We discussed the nonidealities that the traditional class-B oscillators face and reviewed how each structure tries to overcome them. Class-C oscillators improve phase noise for the same power consumption but only when the oscillation amplitude is low enough to keep the core transistors in saturation. Class-D oscillators offer a very low noise without requiring large supply voltages, but they are limited to low supply voltages due to reliability concerns. Class-F oscillators create waveforms with a special ISF that prevents conversion from the circuit thermal noise to phase noise.

References

[1] B. Razavi, "A study of phase noise in CMOS oscillators," *IEEE J. Solid-State Circuits*, vol. 31, no. 3, pp. 331–343, Mar. 1996.

[2] M. Mikhemar, D. Murphy, A. Mirzaei, and H. Darabi, "A cancellation technique for reciprocal-mixing caused by phase noise and spurs," *IEEE J. Solid-State Circuits*, vol. 48, no. 12, pp. 3080–3089, Dec. 2013.

[3] H. Wu, M. Mikhemar, D. Murphy, H. Darabi, and M. F. Chang, "A blocker-tolerant inductor-less wideband receiver with phase and thermal noise cancellation," *IEEE J. Solid-State Circuits*, vol. 50, no. 12, pp. 2948–3024, Dec. 2013.

[4] J. Borremans et al., "A 40 nm CMOS 0.4-6 GHz receiver resilient to out of band blockers," *IEEE J. Solid-State Circuits*, vol. 46, no. 7, pp. 1659–1671, Jul. 2011.

[5] H. Darabi et al., "A quad band GSM/GPRS/EDGE SoC in 65 nm CMOS," *IEEE J. Solid-State Circuits*, vol. 46, no. 4, pp. 870–882, Apr. 2011.

[6] E. J. Baghdady, R. N. Lincoln, and B. D. Nelin, "Short-term frequency stability: Characterization, theory, and measurement," in *Proc. IEEE*, vol. 53, pp. 704–722, Jul. 1965.

[7] L. S. Cutler and C. L. Searle, "Some aspects of the theory and measurement of frequency fluctuations in frequency standards," in *Proc. IEEE*, vol. 54, pp. 136–154, Feb. 1966.

[8] D. B. Leeson, "A simple model of feedback oscillator noises spectrum," in *Proc. IEEE*, vol. 54, pp. 329–330, Feb. 1966.

[9] J. Rutman, "Characterization of phase and frequency instabilities in precision frequency sources; Fifteen years of progress," in *Proc. IEEE*, vol. 66, pp. 1048–1174, Sept. 1978.

[10] A. Hajimiri and T. H. Lee, "A general theory of phase noise in electrical oscillators," *IEEE J. Solid-State Circuits*, vol. 33, no. 2, pp. 179–194, Feb. 1998.

[11] A. Demir, A. Mehrotra, and J. Roychowdhury, "Phase noise in oscillators: a unifying theory and numerical methods for characterization," *IEEE Trans. Circuits Syst. I, Fundam. Theory Appl.*, vol. 47, no. 5, pp. 655–674, May 2000.

[12] D. Murphy, J. J. Rael, and A. A. Abidi, "Phase noise in LC oscillators: A phasor-based analysis of a general result and of loaded Q," *IEEE Trans. Circuits Syst. I, Reg. Papers*, vol. 57, no. 6, pp. 1187–1203, June 2010.

[13] Y. Hu, T. Siriburanon, and R. B. Staszewski, "Intuitive understanding of flicker noise reduction via narrowing of conduction angle in voltage-biased oscillators," *IEEE Trans. on Circuits and Systems II (TCAS-II)*, pp. 1–5, 2019.

[14] A. Hajimiri and T. Lee, "Design issues in CMOS differential LC oscillators," *IEEE J. Solid-State Circuits*, vol. 34, no. 5, pp. 717–724, May 1999.

[15] J. Rael and A. Abidi, "Physical processes of phase noise in differential LC oscillators," *in Proc. IEEE Custom Integr. Circuits Conf.*, Sept. 2000, pp. 569–572.

[16] P. Andreani et al., "A study of phase noise in Colpitts and LC-tank CMOS oscillators," *IEEE J. Solid-State Circuits*, vol. 40, no. 5, pp. 1107–1118, May 2005.

[17] E. Hegazi, H. Sjoland, and A. A. Abidi, "A filtering technique to lower LC oscillator phase noise," *IEEE J. Solid-State Circuits*, vol. 36, no. 12, pp. 1921–1930, Dec. 2001.

[18] A. Hajimiri, "Distributed integrated circuits: An alternative approach to high-frequency design," *IEEE Commun. Mag.*, vol. 40, no. 2, pp. 168–173, Feb. 2002.

[19] H. Chang, X. Cao, U. K. Mishra, and R. A. York, "Phase noise in coupled oscillators: Theory and experiment," *IEEE Trans. Microw. Theory Tech.*, vol. 45, no. 5, pp. 604–615, May 1997.

[20] L. Roman, A. Bonfanti, S. Levantino, C. Samori, and A. L. Lacaita, "5-GHz oscillator array with reduced flicker up-conversion in 0.13-μm CMOS," *IEEE J. Solid-State Circuits*, vol. 41, no. 11, pp. 2457–2467, Nov. 2006.

[21] A. Mazzanti and P. Andreani, "Class-C harmonic CMOS VCOs, with a general result on phase noise," *IEEE J. Solid-State Circuits*, vol. 43, no. 12, pp. 2716–2729, Dec. 2008.

[22] M. Tohidian, A. Fotowat-Ahmadi, M. Kamarei, and F. Ndagijimana, "High-swing class-C VCO,'" in *Proc. IEEE Eur. Solid-State Circuits Conf.*, Sep. 2011, pp. 495–498.

[23] L. Fanori and P. Andreani, "Highly efficient class-C CMOS VCOs, including a comparison with class-B VCOs," *IEEE J. Solid-State Circuits*, vol. 48, no. 7, pp. 1730–1740, Jul. 2013.

[24] W. Deng, K. Okada, and A. Matsuzawa, "A feedback class-C VCO with robust startup condition over PVT variations and enhanced oscillation swing," in *IEEE Eur. Solid-State Circuits Conf. (ESSCIRC)*, Sep. 2011, pp. 499–502.

[25] J. Chen, F. Jonsson, M. Carlsson, C. Hedenas, and L. R. Zheng, "A low power, start-up ensured and constant amplitude class-C VCO in 0.18 μm CMOS," *IEEE Microw. Wireless Compon. Lett.*, vol. 21, no. 8, pp. 427–429, Aug. 2011.

[26] K. Okada, Y. Nomiyana, R. Murakami, and A. Matsuzawa, "A 0.114 mW dual-conduction class-C CMOS VCO with 0.2 V power supply," in *Proc. IEEE Symp. Circuits*, Jun. 2009, pp. 228–229.

[27] L. Fanori, A. Liscidini, and P. Andreani, "A 6.7-to-9.2 GHz 55 nm CMOS hybrid class-B/class-C cellular TX VCO," in *IEEE Int. Solid-State Circuits Conf. (ISSCC) Dig. Tech. Papers*, Feb. 2012, pp. 354–356.

[28] C. Li and A. Liscidini, "Class-C PA-VCO cell for FSK and GFSK transmitters," *IEEE J. Solid-State Circuits*, vol. 51, no. 7, pp. 1537–1546, Jul. 2016.

[29] P. Baxandall, "Transistor sine-wave LC oscillators. Some general considerations and new developments," *Proc. IEE–Part B: Electron. Commun. Eng.*, vol. 106, no. 16, pp. 748–758, May 1959.

[30] L. Fanori and P. Andreani, "A 2.5-to-3.3 GHz CMOS class-D VCO," in *IEEE Int. Solid-State Circuits Conf. (ISSCC) Dig. Tech. Papers*, Feb. 2013, pp. 346–347.

[31] L. Fanori and P. Andreani, "Class-D CMOS oscillators," *IEEE J. Solid-State Circuits*, vol. 48, no. 12, pp. 3105–3119, Dec. 2013.

[32] A. G. Roy, S. Dey, J. B. Goins, T. S. Fiez, and K. Mayaram, "350 mV, 5 GHz class-D enhanced swing differential and quadrature VCOs in 65 nm CMOS," *IEEE J. Solid-State Circuits*, vol. 50, no. 8, pp. 1833–1447, Aug. 2015.

[33] Y. Yoshihara, H. Majima, and R. Fujimoto, "A 0.171 mW, 2.4 GHz class-D VCO with dynamic supply voltage control," in *IEEE Eur. Solid-State Circuits Conf. (ESSCIRC)*, 2014, pp. 339–342.

[34] L. Fanori, T. Mattsson, and P. Andreani, "A class-D CMOS DCO with an on-chip LDO," in *IEEE Eur. Solid-State Circuits Conf. (ESSCIRC)*, 2014, pp. 335–338.

[35] H. Kim, S. Ryu, Y. Chung, J. Choi, and B. Kim, "A low phase-noise CMOS VCO with harmonic tuned LC tank," *IEEE Trans. Microw. Theory Tech.*, vol. 54, no. 7, pp. 2917–2923, Jul. 2006.

[36] D. Manstretta and R. Castello, "An intuitive analysis of phase noise fundamental limits in LC oscillators", in *International Conference on Noise and Fluctuations (ICNF)*, 2015.

[37] M. Babaie, and R. B. Staszewski, "A class-F CMOS oscillator," *IEEE J. Solid-State Circuits*, vol. 48, no. 12, pp. 3120–3133, Dec. 2013.

[38] F. W. Kuo, R. Chen, K. Yen, H. Y. Liao, C. P. Jou, F. L. Hsueh, M. Babaie, and R. B. Staszewski, "A 12 mW all-digital PLL based on class-F DCO for 4G phones in 28 nm CMOS," in *Proceedings of IEEE VLSI Circuits Symposium*, 2014, pp. 1–2.

[39] M. Babaie and R. B. Staszewski, "An ultra-low phase noise class-F_2 CMOS oscillator with 191 dBc/Hz FOM and long term reliability," *IEEE J. Solid-State Circuits*, vol. 50, no. 3, pp. 679–692, Mar. 2015.

3

A Class-F$_3$ CMOS Oscillator

An oscillator topology demonstrating an improved phase noise performance is introduced and analyzed in this chapter. It exploits a time-variant phase noise model with insights into the phase noise conversion mechanisms. This oscillator enforces a pseudo-square voltage waveform around the LC tank by increasing the third harmonic of the fundamental oscillation voltage through an additional impedance peak. This auxiliary impedance peak is realized by a transformer with moderately coupled resonating windings. As a result, the effective impulse sensitivity function (ISF) decreases, thus reducing the oscillator's effective noise factor such that a significant improvement in the oscillator phase noise and power efficiency is achieved. A comprehensive study of circuit-to-phase-noise conversion mechanisms of different oscillators' structures shows that the class-F$_3$ exhibits the lowest phase noise at the same tank's quality factor and supply voltage. The prototype of the class-F$_3$ oscillator is implemented in TSMC 65-nm standard CMOS. It exhibits average phase noise of -142 dBc/Hz at 3 MHz offset from the carrier over 5.9–7.6 GHz tuning range with figure of merit of 192 dBc/Hz. The oscillator occupies 0.12 mm^2 while drawing 12 mA from 1.25 V supply.

3.1 Introduction

Designing voltage-controlled and digitally controlled oscillators (VCO, DCO) of high spectral purity and low power consumption is quite challenging, especially for GSM transmitter (TX), where the oscillator phase noise must be less than -162 dBc/Hz at 20 MHz offset frequency from 915 MHz carrier [1]. At the same time, the RF oscillator consumes disproportionate amount of power of an RF frequency synthesizer [2, 3] and burns more than 30% of the cellular RX power [4, 5]. Consequently, any power reduction of RF oscillators will greatly benefit the overall transceiver power efficiency and

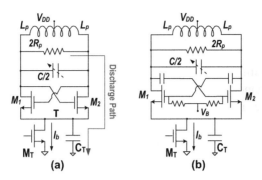

Figure 3.1   Oscillator schematic: (a) traditional class-B; (b) class-C.

ultimately the battery lifetime. This motivation has encouraged an intensive research to improve the power efficiency of an RF oscillator while satisfying the strict phase noise requirements of the cellular standards.

The traditional class-B oscillator (Figure 3.1(a)) is the most prevalent architecture due to its simplicity and robustness. However, as shown in Chapter 2, its phase noise and power efficiency performance drops dramatically by replacing the ideal current source with a real one. For the best performance, the oscillation amplitude should be near supply voltage V_{DD} [6, 7]. Therefore, the gm-devices $M_{1/2}$ enter deep triode for part of the oscillation period. The low impedance path between node "T" due to M_T together with $M_{1/2}$ entering deep triode degrades Q-factor of the tank dramatically and phase noise improvement by increasing oscillation voltage would be negligible.

The noise filtering technique [8] provides a relatively high impedance between the gm-devices and the current source. Hence, the structure maintains the intrinsic Q-factor of the tank during the entire oscillation period. However, it requires an extra resonator sensitive to parasitic capacitances, increasing the design complexity, area, and cost.

As we discussed in Chapter 2, the class-C oscillator (Figure 3.1(b)) prevents the gm-devices from entering the triode region [9, 10]. Hence, the tank Q-factor is preserved throughout the oscillation period. By changing the drain current shape to the "tall and narrow" form for the class-C operation, the oscillator saves 36% power. However, the constraint of avoiding entering the triode region limits the maximum oscillation amplitude of the class-C oscillator to around $V_{DD}/2$, for the case of bias voltage V_B as low as a threshold voltage of the active devices, which limits the lowest achievable phase noise performance.

Harmonic tuning oscillator enforces a pseudo-square voltage waveform around the LC tank through increasing the third harmonic component of the fundamental oscillation voltage through an additional tank impedance peak at that frequency. Kim et al. [11] exploited this technique to improve the phase noise performance of the LC oscillator by increasing the oscillation zero-crossings' slope. However, that structure requires more than two separate LC resonators to make the desired tank input impedance. It increases die area and cost and decreases tuning range due to larger parasitics. Furthermore, the oscillator transconductance loop gain is the same for both resonant frequencies, thus raising the probability of undesired oscillation at the auxiliary tank input impedance. Here, we show how to resolve the concerns and quantify intuitively and theoretically the phase noise and power efficiency improvement of the class-F_3 oscillator compared to other structures [12, 13, 31].

The chapter is organized as follows: Section 3.2 establishes the environment to introduce the class-F_3 oscillator. The circuit-to-phase-noise conversion mechanisms are studied in Section 3.3. Section 3.4 presents extensive measurement results of the prototype, while Section 3.5 wraps up this chapter with conclusions.

3.2 Evolution Towards Class-F₃ Oscillator

Suppose the oscillation voltage around the tank was a square wave instead of a sinusoidal. As a consequence, the oscillator would exploit the special ISF [14] properties of the square-wave oscillation voltage to achieve a better phase noise and power efficiency. However, the gm-devices would work in the triode region (shaded area in Figure 3.2(b)) even longer than in the case

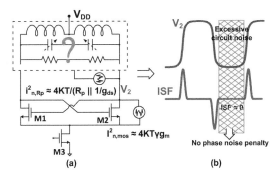

Figure 3.2 LC-tank oscillator: (a) noise sources; (b) targeted oscillation voltage (top) and its expected ISF (bottom).

of the sinusoidal oscillator. Hence, the loaded resonator and gm-device inject more noise to the tank. Nevertheless, ISF value is expected to be negligible in this time span due to the zero derivative of the oscillation voltage [14]. Although the circuit injects huge amount of noise to the tank, the noise cannot change the phase of the oscillation voltage and thus there is no phase noise degradation.

3.2.1 Realizing a Square Wave Across the LC Tank

The above reasoning indicates that the square-wave oscillation voltage has special ISF properties that are beneficial for the oscillator phase noise performance. But how can a square wave be realized across the tank? Let us take a closer look at the traditional oscillator in the frequency domain. As shown in Figure 3.3, the drain current of a typical LC-tank oscillator is approximately a square wave. Hence, it ideally has a fundamental and odd harmonic components. On the other hand, the tank input impedance has a magnitude peak only at the fundamental frequency. Therefore, the tank filters out the harmonic components of the drain current and finally a sinusoidal wave is seen across the tank.

Now, suppose the tank offers another input impedance magnitude peak around the third harmonic of the fundamental frequency (see Figure 3.4). The tank would be prevented from filtering out the third harmonic component of the drain current. Consequently, the oscillation voltage will contain a significant amount of the third harmonic component in addition to the fundamental:

$$V_{in} = V_{p1} \sin\left(\omega_0 t\right) + V_{p3} \sin\left(3\omega_0 t + \Delta\phi\right) \tag{3.1}$$

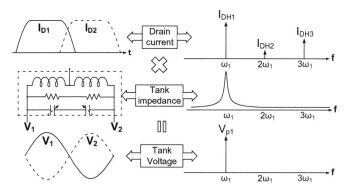

Figure 3.3   Traditional oscillator waveforms in time and frequency domains.

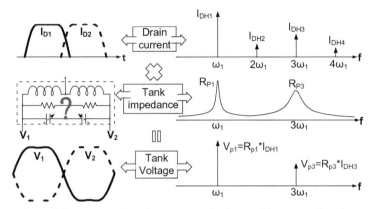

Figure 3.4 New oscillator's waveforms in time and frequency domains.

ζ is defined as the magnitude ratio of the third-to-first harmonic components of the oscillation voltage.

$$\zeta = \frac{V_{p3}}{V_{p1}} = \left(\frac{R_{p3}}{R_{p1}}\right)\left(\frac{I_{DH3}}{I_{DH1}}\right) \approx 0.33 \left(\frac{R_{p3}}{R_{p1}}\right), \tag{3.2}$$

where R_{p1} and R_{p3} are the tank impedance magnitudes at the main resonant frequency ω_1 and $3\omega_1$, respectively. Figure 3.5 illustrates the oscillation voltage and its related expected ISF function (based on the closed-form equation in [14]) for different ζ values. The ISF rms value of the new oscillation waveform can be estimated by the following expression for $-\pi/8 < \Delta\phi < \pi/8$:

$$\Gamma^2_{rms} = \frac{1}{2}\frac{1 + 9\zeta^2}{(1 + 3\zeta)^2}. \tag{3.3}$$

The waveform would become a sinusoidal for the extreme case of $\zeta = 0, \infty$, so (3.3) predicts $\Gamma^2_{rms} = 1/2$, which is well known for the traditional oscillators. Γ^2_{rms} reaches its lowest value of 1/4 for $\zeta = 1/3$, translated to a 3-dB phase noise and FoM improvement compared to the traditional oscillators. Furthermore, ISF of the new oscillator is negligible while the circuit injects significant amount of noise to the tank. Consequently, the oscillator FoM improvement could be larger than that predicted by just the ISF rms reduction.

3.2.2 F₃ Tank

The argumentation related to Figure 3.4 advocates the use of two resonant frequencies with a ratio of 3. The simplest way of realizing that would

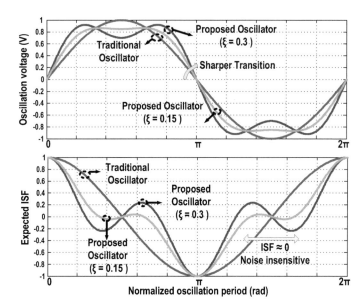

Figure 3.5 The effect of adding third harmonic in the oscillation waveform (top) and its expected ISF (bottom).

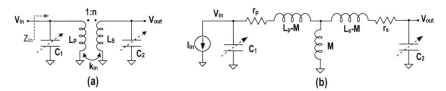

Figure 3.6 Transformer-based resonator (a) and its equivalent circuit (b).

be with two separate inductors [11, 15]. However, this will be bulky and inefficient. The chosen option in this work is a transformer-based resonator. The preferred resonator consists of a transformer with turns ratio n and tuning capacitors C_1 and C_2 at the transformer's primary and secondary windings, respectively (see Figure 3.6). Equation (3.4) expresses the exact mathematical equation of the input impedance of the tank.

$$Z_{in} = \frac{s^3\left(L_pL_sC_2\left(1-k_m^2\right)\right)+s^2(C_2(L_sr_p+L_pr_s))+s(L_p+r_sr_pC_2))+r_p}{\begin{array}{c}s^4\left(L_pL_sC_1C_2\left(1-k_m^2\right)\right)+s^3\left(C_1C_2\left(L_sr_p+L_pr_s\right)\right)+\\ s^2\left(L_pC_1+L_sC_2+r_pr_sC_1C_2\right)+s\left(r_pC_1+r_sC_2\right)+1\end{array}}, \quad (3.4)$$

where k_m is the magnetic coupling factor of the transformer, r_p and r_s model the equivalent series resistance of the primary L_p and secondary L_s

inductances [16]. The denominator of Z_{in} is a fourth-order polynomial for the imperfect coupling factor (i.e., $k_m < 1$). Hence, the tank contains two different conjugate pole pairs, which realize two different resonant frequencies. Consequently, the input impedance has two magnitude peaks at these frequencies. Note that both resonant frequencies can satisfy the Barkhausen criterion with a sufficient loop gain [17]. However, the resulting multi-oscillation behavior is undesired and must be avoided [18]. In our case, it is preferred to see an oscillation at the lower resonant frequency ω_1 and the additional tank impedance at ω_2 is used to make a pseudo-square waveform across the tank. These two possible resonant frequencies can be expressed as

$$\omega_{1,2}^2 = \frac{1 + \left(\frac{L_s C_2}{L_p C_1}\right) \pm \sqrt{1 + \left(\frac{L_s C_2}{L_p C_1}\right)^2 + \left(\frac{L_s C_2}{L_p C_1}\right)(4k_m^2 - 2)}}{2L_s C_2 \left(1 - k_m^2\right)}. \tag{3.5}$$

The following expression offers a good estimation of the main resonant frequency of the tank for $0.5 \le k_m \le 1$.

$$\omega_1^2 = \frac{1}{(L_p C_1 + L_s C_2)} \tag{3.6}$$

However, we are interested in the ratio of resonant frequencies as given by

$$\frac{\omega_2}{\omega_1} = \sqrt{\frac{1 + X + \sqrt{1 + X^2 + X(4k_m^2 - 2)}}{1 + X - \sqrt{1 + X^2 + X(4k_m^2 - 2)}}} \tag{3.7}$$

where X-factor is defined as

$$X = \left(\frac{L_s}{L_p} \cdot \frac{C_2}{C_1}\right). \tag{3.8}$$

Equation (3.7) indicates that the resonant frequency ratio ω_2/ω_1 is just a function of the transformer inductance ratio L_s/L_p, tuning capacitance ratio C_2/C_1, and transformer magnetic coupling factor k_m. The relative matching of capacitors (and inductors) in today's CMOS technology is expected to be much better than 1%, while the magnetic coupling is controlled through lithography that precisely sets the physical dimensions of the transformer. Consequently, the relative position of the resonant frequencies is not sensitive to the process variation. The ω_2/ω_1 ratio is illustrated versus X-factor for different k_m in Figure 3.7. As expected, the ratio moves to higher values for larger k_m and finally the second resonance disappears for the perfect coupling

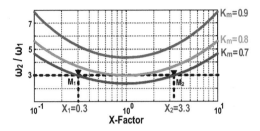

Figure 3.7   Ratio of the tank resonant frequencies versus X-factor for different k_m.

factor. The ratio of ω_2/ω_1 reaches the desired value of 3 at two points for the coupling factor of less than 0.8. Both points put ω_2 at the correct position of $3\omega_1$. However, the desired X-factor should be chosen based on the magnitude ratio R_{p2}/R_{p1} of the tank input impedance at resonance. The sum of the even orders of the denominator in (3.4) is zero at resonant frequencies. It can be shown that the first-order terms of the numerator and the denominator are dominant at ω_1. By using (3.6), assuming $Q_p = L_p\omega/r_p$, $Q_s = L_s\omega/r_s$, the tank input impedance at the fundamental frequency is expressed as

$$R_{p1} \approx \frac{L_p}{\omega_1 \left(\frac{L_p C_1}{Q_p} + \frac{L_s C_2}{Q_s} \right)} \overset{Q_p=Q_s=Q_0}{\Longrightarrow} R_{p1} \approx L_p\omega_1 Q_0. \tag{3.9}$$

On the other hand, it can be shown that the third-order terms of the numerator and the denominator are dominant in (3.4) at $\omega_2 = 3\omega_1$. It follows that

$$R_{p2} \approx \frac{\left(1 - k_m^2\right)}{C_1\omega_2 \left(\frac{1}{Q_p} + \frac{1}{Q_s} \right)} \overset{Q_p=Q_s=Q_0}{\Longrightarrow} R_{p2} \approx \frac{Q_0 \left(1 - k_m^2\right)}{2 \, C_1 \, \omega_2}. \tag{3.10}$$

R_{p2} is a strong function of the coupling factor of the transformer and thus the resulting leakage inductance. Weaker magnetic coupling will result in higher impedance magnitude at ω_2 and, consequently, the second resonance needs a lower transconductance gain to excite. It could even become a dominant pole and the circuit would oscillate at ω_2 instead of ω_1. This phenomenon has been used to extend the oscillator tuning range in [17, 19], and [20]. As explained before, R_{p2}/R_{p1} controls the amount of the third harmonic component of the oscillation voltage. The impedance magnitude ratio is equal to

$$\frac{R_{p2}}{R_{p1}} \approx \frac{\left(1 - k_m^2\right)(1 + X)}{6}. \tag{3.11}$$

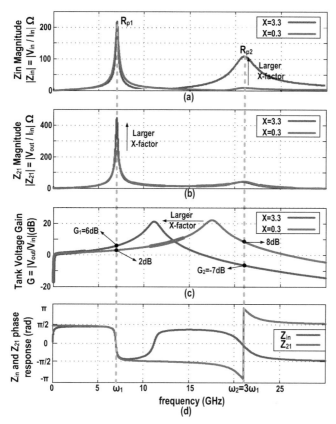

Figure 3.8 The transformer-based tank characteristics: (a) the input impedance, Z_{in} magnitude; (b) the trans-impedance, Z_{21} magnitude; (C) transformer's secondary to primary voltage gain; (d) the phase of Z_{in} and Z_{21} (momentum simulation).

Hence, the smaller X-factor results in lower tank equivalent resistance at $\omega_2 = 3\omega_1$. Thus, the tank filters out more of the third harmonic of the drain current and the oscillation voltage becomes more sinusoidal. Figure 3.8(a) illustrates momentum simulation results of Z_{in} of the transformer-based tank versus frequency for both X-factors that satisfy the resonant frequency ratio of 3. The larger X-factor offers significantly higher tank impedance at ω_2, which is entirely in agreement with the theoretical analysis.

The X-factor is defined as a product of the transformer inductance ratio L_s/L_p and tuning capacitance ratio C_2/C_1. This leads to a question of how to best divide X-factor between the inductance and capacitance ratios. In general, larger L_s/L_p results in higher inter-winding voltage gain, which translates to sharper transition at zero-crossings and larger oscillation amplitude at the secondary winding. Both of these effects have a direct consequence on the phase noise improvement. However, the transformer Q-factor drops by increasing the turns ratio. In addition, very large oscillation voltage swing brings up reliability issues due to the gate-oxide breakdown. It turns out that the turns ratio of 2 can satisfy the aforementioned constraints altogether.

3.2.3 Voltage Gain of the Tank

The transformer-based resonator, whose schematic was shown in Figure 3.6, offers a filtering function on the signal path from the primary to the secondary windings. The tank voltage gain is derived as

$$G\left(s\right) = \frac{V_{out}}{V_{in}} = \frac{Ms}{s^3(L_pL_sC_2(1-k_m^2))+s^2(C_2(L_sr_p+L_pr_s))+s(L_p+r_sr_pC_2))+r_p}.$$
(3.12)

Bode diagram of the tank voltage gain transfer function is shown in Figure 3.9. The tank exhibits a 20 dB/dec attenuation for frequencies lower than the first pole and offers a constant voltage gain at frequencies between the first pole and the complex conjugate pole pair at ω_p. The gain plot reveals an interesting peak at frequencies around ω_p, beyond which the filter gain drops at the -40 dB/dec slope. The low frequency pole is estimated by

$$\omega_{p1} = \frac{r_p}{L_p}.$$
(3.13)

By substituting $r_p = L_p\omega/Q_p$, $r_s = L_s\omega/Q_s$ and assuming $Q_p \cdot Q_s \gg 1$, the tank gain transfer function can be simplified to the following equation for

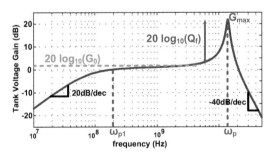

Figure 3.9 Typical secondary-to-primary winding voltage gain of the transformer-based resonator versus frequency.

the frequencies beyond ω_{p1}:

$$G(s) = \frac{\frac{M}{L_p}}{s^2 \left(L_s C_2 \left(1 - k_m^2\right)\right) + s \left(L_s C_2 \omega \left(\frac{1}{Q_p} + \frac{1}{Q_s}\right)\right) + 1}. \quad (3.14)$$

The main characteristics of the tank voltage gain can be specified by considering it as a biquad filter.

$$G(s) = \frac{G_0}{\left(\frac{s}{\omega_p}\right)^2 + \left(\frac{s}{\omega_p Q_f}\right) + 1}, \quad (3.15)$$

where

$$G_0 = k_m n. \quad (3.16)$$

The peak frequency is estimated by

$$\omega_p = \sqrt{\frac{1}{L_s C_2 \left(1 - k_m^2\right)}} \quad (3.17)$$

Q_f represents the amount of gain jump around ω_p and expressed by

$$Q_f = \frac{\left(1 - k_m^2\right)}{\frac{1}{Q_p} + \frac{1}{Q_s}}. \quad (3.18)$$

Hence, the maximum voltage gain is calculated by

$$G_{max} = k_m n \times \frac{\left(1 - k_m^2\right)}{\frac{1}{Q_p} + \frac{1}{Q_s}}. \quad (3.19)$$

Equation (3.19) and Figure 3.9 demonstrate that the transformer-based resonator can offer the voltage gain above $k_m n$ at the frequencies near ω_p for $k_m < 1$ and the peak magnitude is increased by improving Q-factor of the transformer individual inductors. Consequently, ω_1 should be close to ω_p to have higher passive gain at the fundamental frequency and more attenuation at its harmonic components. Equations (3.6) and (3.17) indicate that ω_p is always located at frequencies above ω_1 and the frequency gap between them decreases with greater X-factor. Figure 3.8(c) illustrates the voltage gain of the transformer-based tank for two different X-factors that exhibit the same resonant frequencies. The transformer peak gain happens at much higher

Table 3.1 Normalized zero-crossing slope of the novel oscillator

	Normalized Zero-crossing Slope
Traditional LC	1
Novel tank (primary)	$1 + 3\zeta = 1 + 3 \cdot 1/6 = 1.5$
Novel tank (secondary)	$G_1\text{-}3G_2\zeta = 2.1 - 3 \cdot 0.4 \cdot 1/6 = 1.9$

frequencies for the smaller X-factor and, therefore, the gain is limited to only $k_m n$ (2 dB in this case) at ω_1. However, X-factor is around 3 for the new oscillator and, as a consequence, ω_p moves lower and much closer to ω_1. Now, the tank offers higher voltage gain ($G_1 = 6$ dB in this case) at the main resonance and more attenuation ($G_2 = -7$ dB) at ω_2. This former translates to larger oscillation voltage swing and thus better phase noise.

As can be seen in Figure 3.8(d), the input impedance Z_{in} phase is zero at the first and second resonant frequencies. Hence, any injected third harmonic current has a constructive effect resulting in sharper zero-crossings and flat peak for the transformer's primary winding voltage. However, the tank trans-impedance, Z_{21} phase shows a 180 degree phase difference at ω_1 and $\omega_2 = 3\omega_1$. Consequently, the third harmonic current injection at the primary windings leads to a slower zero-crossings slope at the transformer's secondary, which has an adverse outcome on the phase noise performance of the oscillator. Figure 3.8(a–c) illustrates that this transformer-based resonator effectively filters out the third harmonic component of the drain current at the secondary winding in order to minimize these side effects and zero-crossings are sharpened by tank's voltage gain (G_1) at ω_1. Table 3.1 shows that the zero-crossing slope of this oscillator at both transformer's windings are improved compared to the traditional oscillator for the same V_{DD}, which is translated to shorter commutating time and lower active device noise factor.

3.2.4 Class-F₃ Oscillator

The desired tank impedance, inductance, and capacitance ratios were determined above to enforce the pseudo-square-wave oscillation voltage around the tank. Now, two transistors should be customarily added to the transformer-based resonator to sustain the oscillation. There are two options, however, as shown in Figure 3.10, for connecting the transformer to the active gm-devices. The first option is a transformer-coupled class-F₃ oscillator in which the secondary winding is connected to the gate of the gm-devices. The second option is a cross-coupled class-F₃ oscillator with a floating secondary transformer winding, which only physically connects to tuning capacitors C_2.

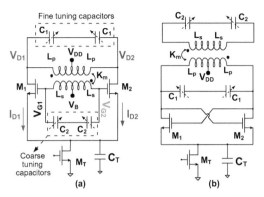

Figure 3.10 Two options of the transformer-based class-F_3 oscillator: (a) transformer-coupled and (b) cross-coupled. The first option was chosen as more advantageous in this work.

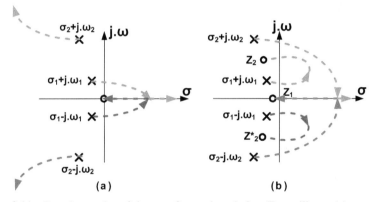

Figure 3.11 Root-locus plot of the transformer-based class-F_3 oscillator: (a) transformer-coupled structure of Figure 3.10(a); and (b) cross-coupled structure of Figure 3.10(b).

The oscillation voltage swing, the equivalent resonator quality factor, and tank input impedance are the same for both options. However, the gm-device sustains larger voltage swing in the first option. Consequently, its commutation time is shorter and the active device noise factor is lower. In addition, the gm-device generates higher amount of the third harmonic, which results in sharper pseudo-square oscillation voltage with lower ISF rms value. The second major difference is about the possibility of oscillation at ω_2 instead of ω_1. The root-locus plot in Figure 3.11 illustrates the route of pole movements towards zeros for different values of the oscillator loop transconductance gain (G_m). As can be seen in Figure 3.11(b), both resonant frequencies (ω_1, ω_2)

can be excited simultaneously with a relatively high value of G_m for the cross-coupled class-F$_3$ oscillator of Figure 3.10(b). It can increase the likelihood of the undesired oscillation at ω_2. However, the transformer-coupled circuit of Figure 3.10(a) demonstrates a different behavior. The lower frequency conjugate pole pair moves into the right-hand plane by increasing the absolute value of G_m, while the higher poles are pushed far away from imaginary axis (see Figure 3.11(a)). This guarantees that the oscillation can only happen at ω_1. Consequently, it becomes clear that the transformer-coupled oscillator is a better option due to its phase noise performance and the guaranty of operation at the right resonant frequency. Nevertheless, the gate parasitic capacitance appears at the drain through a scaling factor of n^2, which reduces its tuning range somewhat as compared to the cross-coupled candidate.

Figure 3.12(a) illustrates the unconventional oscillation voltage waveforms of this transformer-coupled class-F$_3$ oscillator. As specified in Section 3.2.3, the third harmonic component of the drain voltage attenuates at the gate and thus a sinusoidal wave is seen there. The gate–drain voltage swing goes as high as $2.7 \cdot V_{DD}$ due to the significant voltage gain of the tank. Hence, using thick-oxide gm-devices is a constraint to satisfy the time-dependent dielectric breakdown (TDDB) issue for less than 0.01% failure rate during 10 years of the oscillator operation [21, 22]. The costs are larger parasitics capacitance and slightly lower frequency tuning range.

The frequency tuning requires a bit different consideration in the class-F$_3$ oscillator. Both C_1 and C_2 must, at a coarse level, be changed simultaneously to maintain $L_s C_2 / L_p C_1$ ratio such that ω_2 aligns with $3\omega_1$.

Figure 3.12(b) shows the transient response of the class-F oscillator. At power up, the oscillation voltage is very small and the drain current pulses have narrow and tall shape. Even though the tank has an additional impedance

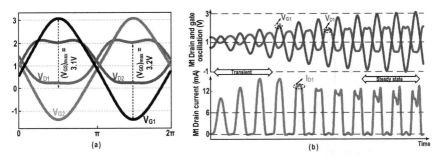

Figure 3.12   (a) Oscillation voltage waveforms and (b) transient response of the class-F$_3$ oscillator.

at $3\omega_1$, the third harmonic component of the drain current is negligible and, consequently, the drain oscillation resembles a sinusoid. At steady state, gate oscillation voltage swing is large and the gm-device drain current is square wave. Consequently, the combination of the tank input impedance with significant drain's third harmonic component results in the pseudo-square-wave for the drain oscillation voltage. This justifies its "class-F₃" designation.

3.3 Class-F₃ Phase Noise Performance

3.3.1 Quality Factor of Transformer-Based Resonator

The Q-factor of the complex tank, which comprises two coupled resonators, does not appear to be as straightforward in intuitive understanding as the Q-factor of the individual physical inductors. It is, therefore, imperative to understand the relationship between the open-loop Q-factor of the tank versus the Q-factor of the inductive and capacitive parts of the resonator.

First, suppose the tuning capacitance losses are negligible. Consequently, the oscillator equivalent Q-factor just includes the tank's inductive part losses. The open-loop Q-factor of the oscillator is defined as $\omega_0/2 \cdot d\phi/dw$, where ω_0 is the resonant frequency and $d\phi/dw$ denotes the slope of the phase of the oscillator open-loop transfer function [23]. To determine the open-loop Q, we need to break the oscillator loop at the gate of M_1, as shown in Figure 3.13. The open-loop transfer function is thus given by

$$H\left(s\right) = \frac{V_{out}}{I_{in}} = \frac{Ms}{As^4 + Bs^3 + Cs^2 + Ds + 1}, \qquad (3.20)$$

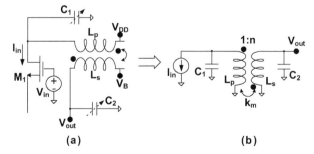

(a)   (b)

Figure 3.13 Open-loop circuit for unloaded Q-factor calculation (a); its equivalent circuit (b).

where $A = L_pL_sC_1C_2\left(1 - k_m^2\right)$, $B = C_1C_2\left(L_sr_p + L_pr_s\right)$, $C = L_pC_1 + L_sC_2 + r_pr_sC_1C_2$, and $D = r_pC_1 + r_sC_2$. After carrying out lengthy algebra and considering $\left(1 - C\omega^2 + A\omega^4 \approx 0\right)$ at the resonant frequencies,

$$Q_i = -\frac{\omega}{2}\frac{d\phi\left(\omega\right)}{d\omega} = \frac{\left(C\omega - 2A\omega^3\right)}{\left(D - B\omega^2\right)}. \tag{3.21}$$

Substituting A, B, C, and D into (3.21), then swapping r_p and r_s with $L_p\omega/Q_p$ and $L_s\omega/Q_s$, respectively, and assuming $Q_pQ_s \gg 1$, we obtain

$$Q_i = \frac{\left(L_pC_1 + L_sC_2\right) - 2\left(L_pL_sC_1C_2\left(1 - k_m^2\right)\right)\omega^2}{\left(\frac{L_pC_1}{Q_p} + \frac{L_sC_2}{Q_s}\right) - \left(C_1C_2L_sL_p\left(\frac{1}{Q_p} + \frac{1}{Q_s}\right)\right)\omega^2}. \tag{3.22}$$

Substituting (3.5) as ω into the above equation and carrying out the mathematics, the tank's inductive part Q-factor at the main resonance is

$$Q_i = \frac{\left(1 + X^2 + 2k_mX\right)}{\left(\frac{1}{Q_p} + \frac{X^2}{Q_s}\right)}. \tag{3.23}$$

To help with an intuitive understanding, let us consider a boundary case. Suppose that C_2 is negligible. Therefore, X-factor is zero and (3.23) predicts that the Q_i equals to Q_p. This is not surprising because no energy would be stored at the transformer's secondary winding and its Q-factor would not have any contribution to the equivalent Q-factor of the tank. In addition, (3.23) predicts that the equivalent Q-factor of the tank's inductive part can exceed Q-factors of the individual inductors. This clearly proves Q-factor enhancement over that of the transformer's individual inductors. The maximum tank's inductive part Q-factor is obtained at the following X-factor for a given k_m, Q_p, and Q_s.

$$X_{Qmax} = \frac{Q_s}{Q_p}. \tag{3.24}$$

For a typical case of $Q_s = Q_p = Q_0$, the maximum Q_i at ω_1 is calculated by

$$X_{Qi,max} = 1 \rightarrow Q_{i,max} = Q_0\left(1 + k_m\right). \tag{3.25}$$

The above equation indicates that the equivalent Q-factor of the inductive part of the transformer-based resonator can be enhanced by a factor of $1 + k_m$ at the optimum state. However, it does not necessarily mean that the Q-factor of the transformer-based tank generally is superior to the simple LC resonator. The reason is that it is not possible to optimize the Q-factor of both windings of a 1:n transformer at a given frequency and one needs to use lower metal layers for the transformer cross connections, which results in more losses and lower Q-factor [24, 25]. For this prototype, the X-factor is around 3 with $k_m = 0.7$ and the simulated Q_p and Q_s are 14 and 20, respectively. Based on (3.23), the equivalent Q-factor of the inductive part of the tank would be about 26, which is higher than that of the transformers' individual inductors. The Q-factor of the switched capacitance largely depends on the tuning range (TR) and operating frequency of the oscillator and is about 42 for the TR of 25% at 7 GHz resulting in an average Q-factor of 16 for the tank in this design.

3.3.2 Phase Noise Mechanism in Class-F₃ Oscillator

According to the linear time-variant model [14], the phase noise of the oscillator at an offset frequency $\Delta\omega$ from its fundamental frequency is expressed as

$$L(\Delta\omega) = 10 \log_{10} \left(\frac{\sum_i N_{L,i}}{2\, q_{max}^2\, (\Delta\omega)^2} \right), \qquad (3.26)$$

where q_{max} is the maximum charge displacement across the tuning capacitor C and $N_{L,i}$ is the effective noise produced by ith device given by

$$N_{L,i} = \frac{1}{2\pi N^2} \int_0^{2\pi} \Gamma_i^2(t)\, \overline{i_{n,i}^2(t)} dt \qquad (3.27)$$

where $\overline{i_{n,i}^2(t)}$ is the white current noise power density of the ith noise source, Γ_i is its relevant ISF function from the corresponding ith device noise, and N is the number of resonators in the oscillator. N is considered one for single-ended and two for differential oscillator topologies with a single LC tank [7].

Figure 3.14 illustrates the major noise sources of CMOS class-B, class C, and class-F₃ oscillators. R_p and $G_{ds1,2}(t)$ represent the equivalent tank parallel resistance and channel conductance of the gm transistors, respectively. On the other hand, $G_{m1,2}$ and G_{mT} model the noise due to transconductance gain of active core and current source transistors, respectively. By substituting (3.27) into (3.26) and carrying out algebra, the phase noise equation is simplified to

$$L(\Delta\omega) = 10 \log_{10} \left(\frac{K_B T R_p}{2\, Q_t^2\, V_p^2} \cdot F \cdot \left(\frac{\omega_0}{\Delta\omega} \right)^2 \right), \qquad (3.28)$$

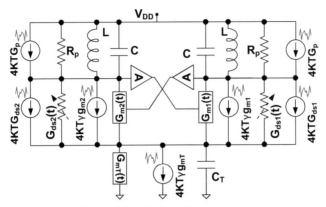

Figure 3.14   RF CMOS oscillator noise sources.

where Q_t is the tank's equivalent quality factor and V_p is the maximum oscillation voltage, derived by

$$V_p = \begin{cases} \left(\dfrac{1}{3}+\zeta\right)\sqrt{\left(1+\dfrac{1}{3\zeta}\right)} \cdot \alpha_I \cdot R_p \cdot I_B, & \dfrac{1}{9} \leq \zeta \leq 1 \\[2ex] (1-\zeta) \cdot \alpha_I \cdot R_p \cdot I_B, & 0 \leq \zeta \leq \dfrac{1}{9}, \end{cases} \tag{3.29}$$

where α_I is the current conversion efficiency of the oscillator, expressed as the ratio of the fundamental component of gm-devices drain current to dc current I_B of the oscillator. F in (3.28) is the effective noise factor of the oscillator, expressed by

$$F = \sum_i \frac{1}{2\pi} \int_0^{2\pi} \Gamma_i^2(t) \frac{\overline{i_{n,i}^2(t)} R_p}{4K_B T} dt. \tag{3.30}$$

Suppose that C_T is large enough to filter out the thermal noise of the tail transistor. Consequently, F consists of the noise factor of the tank (F_{tank}), transistor channel conductance (F_{GDS}), and gm of core devices (F_{GM}). The expressions of F_{tank} and F_{GDS} are

$$F_{tank} = \frac{1}{\pi} \int_0^{2\pi} \Gamma_{tank}^2(t) dt = 2\Gamma_{rms}^2 \approx \frac{1+9\zeta^2}{(1+3\zeta)^2} \tag{3.31}$$

$$F_{GDS} = \frac{1}{\pi} \int_0^{2\pi} \Gamma_{MOS}^2(t) G_{DS1}(t) R_p dt \approx 2\Gamma_{rms}^2 R_P \cdot G_{DS1EF}, \tag{3.32}$$

where G_{DSEF1} is the effective drain–source conductance of one of the gm-devices expressed by

$$G_{DS1EF} = G_{DS1}[0] - G_{DS1}[2], \qquad (3.33)$$

where $G_{DS1}[k]$ describes the kth Fourier coefficient of the instantaneous conductance, $G_{ds1}(t)$ [26]. F_{GM} can be calculated by

$$F_{GM} = \frac{1}{\pi} \int_0^{2\pi} \Gamma_{MOS}^2(t) \gamma G_{m1}(t) R_P dt \approx 2\Gamma_{rms}^2 \cdot \gamma \cdot R_P \cdot G_{M1EF}. \quad (3.34)$$

Now, the effective negative transconductance of the oscillator needs to overcome the tank and its own channel resistance losses and, therefore, the noise due to G_M also increases.

$$G_{M1EF} = \frac{1}{A}\left(\frac{1}{R_p} + G_{DS1EF}\right), \qquad (3.35)$$

where A is the voltage gain of feedback path between the tank and MOS gate. By substituting (3.35) into (3.34)

$$F_{GM} = 2\,\Gamma_{rms}^2 \cdot \frac{\gamma}{A} \cdot (1 + R_P G_{DS1EF}). \qquad (3.36)$$

Consequently, the effective noise factor of the oscillator is given by

$$F = 2\,\Gamma_{rms}^2 \cdot \left(1 + \frac{\gamma}{A}\right) \cdot (1 + R_P G_{DS1EF}). \qquad (3.37)$$

This is a general result and is applicable to the class-B, class-C, and class-F$_3$. The oscillator FoM normalizes the phase noise performance to the oscillation frequency and power consumption, yielding

$$FoM = -10 \cdot log_{10}\left(\frac{10^3\,K_B\,T}{2\,Q_t^2\,\alpha_I\,\alpha_V} \cdot 2\Gamma_{rms}^2 \cdot \left(1 + \frac{\gamma}{A}\right) \cdot (1 + R_P G_{DS1EF})\right), \tag{3.38}$$

where α_V is the voltage efficiency, defined as V_P/V_{DD}.

To get a better insight, the circuit-to-phase-noise mechanism, relative phase noise, and power efficiency of different oscillator classes are also investigated and compared together in this section. Figure 3.15(a–f) shows the oscillation voltage and drain current for the traditional, class-C and class-F oscillators for the same V_{DD} (i.e., 1.2 V), tank Q-factor (i.e., 15), and R_P (i.e., 220 Ω).

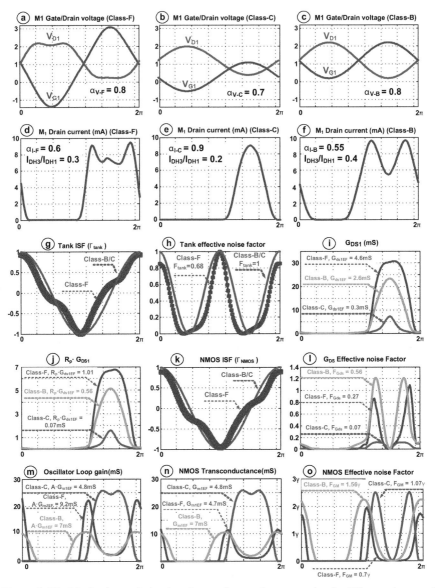

Figure 3.15   Mechanisms of circuit noise to phase noise conversion in different classes of RF CMOS oscillator.

The α_V must be around 0.8 for the class-B and class-F$_3$ oscillators due to the voltage drop V_{dsat} across the tail transistor needed to keep it in saturation. The combination of the tail capacitance and entering the gm-devices into the linear region reduces α_I of class-B from the theoretical value of $2/\pi$ to around 0.55. Fortunately, α_I is maintained around $2/\pi$ for class-F$_3$ due to the pseudo-square drain voltage and larger gate amplitude. The class-C oscillator with a dynamic bias of the active transistor offers significant improvements over the traditional class-C and maximizes the oscillation amplitude without compromising the robustness of the oscillator start-up [27]. Nevertheless, its α_V is around 0.7 to avoid gm-devices entering the triode region. Class-C drain current composed of tall and narrow pulses results in α_I equal to 0.9 (ideally 1).

Obtaining the ISF function is the first step in the calculation of the oscillator's effective noise factor. The class-B/C ISF function is a sinusoid in quadrature with the tank voltage [7, 28]. However, finding the exact equation of class-F$_3$ ISF is not possible; hence, we had to resort to painstakingly long Cadence^{TM} simulations to obtain the ISF curves. Figure 3.15(g) shows the simulated class-F tank equivalent ISF function, which is smaller than the other classes for almost the entire oscillation period.

Figure 3.15(h) demonstrates the tank effective noise factor along the oscillation period for different oscillator classes. The F_{RP} is 32% lower for this class-F$_3$ due to its special ISF properties. The gm-device M_1 channel conductance across the oscillation period is shown in Figure 3.15(i). As expected, $G_{DS1}(t)$ of class-F$_3$ exhibits the largest peak due to high oscillation swing at the gate and, consequently, injects more noise than other structures to the tank. On the other hand, class-C operates only in the saturation region and its effective transistor conductance is negligible. Figure 3.15(j) strongly emphasizes that the gm-device resistive channel noise could even be 7 times higher than the tank noise when the M_1 operates in the linear region. To get a better insight, one need to simultaneously focus on Figures 3.15(j) and (k). Although the class-F$_3$ G_{DS1} generates lots of noise in the second half of the period, its relevant ISF value is very small there. Hence, the excessive transistor channel noise cannot convert to the phase noise and as shown in Figure 3.15(l), the F_{GDS} of class-F$_3$ is one-half of the traditional oscillator. The transconductance loop gains of the different oscillator structures are shown in Figure 3.15(m). Class-F$_3$ needs to exhibit the highest effective transconductance loop gain to compensate its larger gm-devices channel resistance losses. However, half of the required loop gain is covered by the transformer-based tank voltage gain. Figure 3.15(o) demonstrates the active device effective noise factor along the oscillation period. Class-F$_3$ offers the lowest F_{GM} due to its special ISF nature and the passive voltage gain between the tank and gate of the gm-transistors.

Table 3.2 Comparison of different oscillator's classes for the same V_{DD} (1.2 V), tank Q-factor (15), R_P (i.e. 220 Ω), and carrier frequency (7 GHz) at 3 MHz offset frequency

	Theoretical Expression	Class-B	Class-C	Class-F$_3$
F_{RP}	$2\Gamma_{rms}^2$	1 (average)	1 (average)	0.7 (best)
F_{GDS}	$2\Gamma_{rms}^2 R_P G_{DSEF1}$	0.56 (worst)	0.07 (best)	0.27 (average)
F_{GM}	$2\Gamma_{rms}^2 \frac{\gamma}{A}\left(1+R_P G_{DS1EF}\right)$	1.56γ (worst)	1.07γ (average)	0.7γ (best)
F	$2\,\Gamma_{rms}^2\left(1+\frac{\gamma}{A}\right)\left(1+R_P G_{DS1EF}\right)$	5.5 dB (worst)	3.9 dB (average)	2.8 dB (best)
α_I	I_{H1}/I_B	0.55 (worst)	0.9 (best)	0.63 (average)
α_V	V_p/V_{DD}	0.8 (best)	0.7 (average)	0.8 (best)
$PN(dBc/Hz)$	$10\log_{10}\left(\frac{K_B\,T\,R_p}{2\,Q_0^2\,V_p^2}\cdot F\cdot\left(\frac{\omega_0}{\Delta\omega}\right)^2\right)$	-133.5 (worst)	-134 (average)	-136 (best)
$FoM(dB)$	$-10\log_{10}\left(\frac{1000\,K_B T}{2Q_0^2\alpha_I\alpha_V}F\right)$	191.2 (worst)	194.5 (best)	194.2 ($\approx$ best)

Table 3.2 summarizes the performance of different oscillator classes of this example. It can be concluded that class-F$_3$ oscillator achieves the lowest circuit-to-phase-noise conversion along the best phase noise performance with almost the same power efficiency as the class-C oscillator.

The use of transformer in the class-F$_3$ configuration offers an additional reduction of the $1/f^3$ phase noise corner. The transformer inherently rejects the common-mode signals. Hence, the $1/f$ noise of the tail current source can appear at the transformer's primary, but it will be effectively filtered out on the path to the secondary winding. Consequently, the AM-to-PM conversion at the C_2 switched capacitors is entirely avoided.

3.3.3 Class-F$_3$ Operation Robustness

Figure 3.16(a) illustrates the tank input impedance magnitude and phase for the imperfect position of the second resonance frequency ω_2. A 6% mismatch is applied to the C_2/C_1 ratio, which shifts ω_2 to frequencies higher than $3\omega_1$. Hence, the third harmonic of the drain current is multiplied by a lower impedance magnitude with a phase shift resulting in a distorted pseudo-square oscillation waveform as shown in Figure 3.16(b). Intuitively, if the Q-factor at ω_2 was smaller, the tank impedance bandwidth around it would be wider. Therefore, the tank input impedance phase shift and magnitude reduction would be less for a given ω_2 drift from $3\omega_1$. As a consequence, the oscillator would be less sensitive to the position of ω_2 and thus the tuning capacitance ratio. Based on the open-loop Q-factor analysis, substituting $\omega^2 \approx 9/(L_sC_2 + L_pC_1)$ into (3.22), the Q_i is obtained as $0.3Q_0$ at ω_2. Fortunately enough, the proposed tank configuration automatically reduces

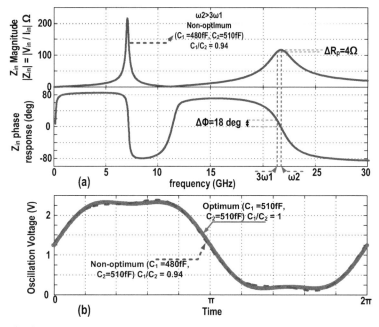

Figure 3.16 Sensitivity of class-F$_3$ oscillator to the position of the second resonant frequency: tank's input impedance magnitude and phase (top); oscillation waveform (bottom).

the equivalent tank Q-factor at ω_2 to 30% of the main resonance Q-factor. This is completely in line with the desire to reduce the sensitivity to the position of ω_2 in class-F$_3$. Consequently, a realistic example ±30 fF variation in C$_1$ from its optimum point has absolutely no major side effects on the oscillator waveform and thus its phase noise performance, as apparent from Figure 3.16. It is strongly emphasized that the circuit oscillates based on ω_1 resonance, and low Q-factor at ω_2 has no adverse consequence on the oscillator phase noise performance.

3.4 Experimental Results

3.4.1 Implementation Details

The class-F$_3$ oscillator, whose schematic was shown in Figure 3.10(a), has been realized in TSMC 1P7M 65-nm CMOS technology with Alucap layer. The differential transistors are thick-oxide devices of 12(4-μm/0.28-μm) dimension to withstand large gate voltage swing. However, the tail current source M$_T$ is implemented as a thin-oxide 500-μm/0.24-μm device biased in saturation. The large channel length is selected to minimize its 1/f noise. Its large drain–bulk and drain–gate parasitic capacitances combined with $C_T = 2$ pF MOM capacitor shunt the M_T thermal noise to ground. The step-up 1:2 transformer is realized by stacking the 1.45 μm Alucap layer on top of the 3.4 μm thick top (M7 layer) copper metal. Its primary and secondary differential self-inductances are about 500 and 1500 pH, respectively, with the magnetic coupling factor of 0.73. The transformer was designed with a goal of maximizing Q-factor of the secondary winding, Q_s, at the desired operating frequency. Based on (3.23), Q_s is the dominant factor in the tank equivalent Q-factor expression, provided $(L_sC_2)/(L_pC_1)$ is larger than one, which is valid for this oscillator prototype. In addition, the oscillation voltage is sinusoidal across the secondary winding. It means the oscillator phase noise is more sensitive to the circuit noise at the secondary winding compared to the primary side with the pseudo-square waveform. Four switched MOM capacitors $B_{C0} - B_{C3}$ placed across the secondary winding realize coarse tuning bits, while the fine control bits $B_{F0} - B_{F3}$ with LSB size of 20 fF adjust the position of ω_2 near $3\omega_1$. The center tap of the secondary winding is connected to the bias voltage, which is fixed around 1 V to guarantee safe oscillator start-up in all process corners. A resistive shunt buffer interfaces the oscillator output to the dynamic divider [2]. A differential output buffer drives a 50-Ω load. The separation of the oscillator core and divider/output buffer voltage supplies and grounds serves to maximize the isolation between the circuit blocks. The die micrograph is shown in Figure 3.17. The oscillator core die area is 0.12 mm^2.

3.4.2 Measurement Results

The measured phase noise at 3.7 GHz (after the on-chip ÷2 divider) at 1.25 V and 12 mA current consumption is shown in Figure 3.18. The phase noise of -142.2 dBc/Hz at 3 MHz offset lies on the 20 dB/dec slope, which extrapolates to -158.7 dBc/Hz at 20 MHz offset (-170.8 dBc/Hz when normalized to 915 MHz) and meets the GSM TX mobile station (MS) specification with a very wide 8 dB margin. The oscillation purity of the class-F$_3$ oscillator is good enough to compare its performance to cellular basestation (BTS) phase noise requirements. The GSM/DCS "Micro" BTS phase noise requirements are easily met. However, the phase noise would be off by 3 dB for the toughest DCS-1800 "Normal" BTS specification at 800 kHz offset frequency [29]. The 1/f^3 phase noise corner is around 700 kHz at the highest frequency due to the asymmetric layout of the oscillator differential nodes further magnified

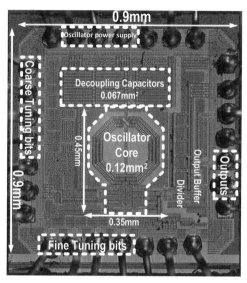

Figure 3.17 Die photograph of class-F$_3$ oscillator.

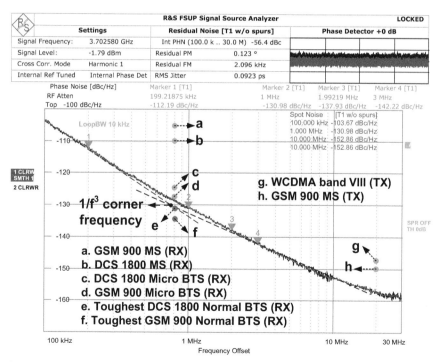

Figure 3.18 Measured phase noise at 3.7 GHz and power dissipation of 15 mW. Specifications (MS: mobile station; BTS: basestation) are normalized to the carrier frequency.

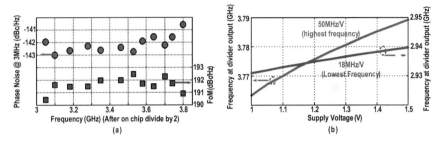

Figure 3.19 (a) Phase noise and figure of merit (FoM) at 3 MHz offset versus carrier frequency and (b) frequency pushing due to supply voltage variation.

by the dominance of parasitics in the equivalent tank capacitance. The $1/f^3$ phase noise corner moves to around 300 kHz at the middle and low parts of the tuning range. The noise floor is -160 dBc/Hz and dominated by thermal noise from the divider and buffers. The oscillator has a 25% tuning range from 5.9 to 7.6 GHz. Figure 3.19(a) shows the average phase noise performance of four samples at 3 MHz offset frequency across the tuning range (after the divider), together with the corresponding FoM. The average FoM is as high as 192 dBc/Hz and varies about 2 dB across the tuning range. The divided output frequency versus supply is shown in Figure 3.19(b) and reveals very low frequency pushing of 50 and 18 MHz/V at the highest and lowest frequencies, respectively.

The phase noise of the class-F₃ oscillator was measured at the fixed frequency of 3.5 GHz for two configurations. In the first configuration, the C_2/C_1 ratio was set to one to align the second resonant frequency ω_2 exactly at the third harmonic of the fundamental frequency ω_1. This is the optimum configuration of the class-F₃ oscillator (Figure 3.20, top). In the second configuration, the oscillation frequency is kept fixed, but an unrealistically high 40% mismatch was applied to the C_2/C_1 ratio, which lowers ω_2, in order to see its effects on the phase noise performance (see Figure 3.20, bottom). As a consequence, the third harmonic component of the drain oscillation voltage is reduced and a phase shift can be seen between voltage waveform components at $3\omega_1$ and ω_1. Therefore, its ISF rms value is worse than optimum, thus causing a 2-dB phase noise degradation in the 20-dB/dec region. In addition, the voltage waveform demonstrates more asymmetry in the rise and fall times, which translates to the non-zero ISF dc value and increases the upconversion factor of the 1/f noise corner of gm-devices. As can be seen in Figure 3.20, the $1/f^3$ phase noise corner is increased by 25% or 100 kHz in the non-optimum case. It results in a 3-dB phase noise penalty in the flicker noise region.

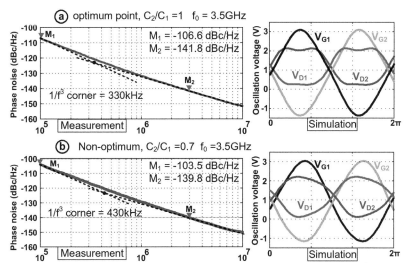

Figure 3.20 Measured phase noise at 3.5 GHz and simulated oscillation waveforms: (a) optimum case; (b) exaggerated non-optimum case.

Table 3.3 summarizes performance of this class-F_3 oscillator and compares it with the relevant oscillators. The class-F_3 demonstrates a 5-dB phase noise and 7-dB FoM improvements over the traditional commercial oscillator [2] with almost the same tuning range. For the same phase noise performance range (-154 to -155 dBc/Hz) at 3-MHz offset for the normalized 915-MHz carrier, the class-F_3 oscillator consumes only 15 mW, which is much lower than that with Colpitts [30], class B/C [10], and clip-and-restore [29] topologies. Only the noise-filtering-technique oscillator [8] offers a better power efficiency but at the cost of an extra dedicated inductor and thus larger die. Also, it uses a 2.5-V supply, thus making it unrealistic in today's scaled CMOS. From the FoM point of view, the class-C oscillator [9] exhibits a better performance than the class-F_3 oscillator. However, the voltage swing constraint in class-C limits its phase noise performance. As can be seen, the class-F_3 demonstrates more than 6 dB better phase noise with almost the same supply voltage. Consequently, the class-F_3 oscillator has reached the best phase noise performance with the highest power efficiency at low voltage supply without the die area penalty of the noise-filtering technique or voltage swing constraint of the class-C VCOs.

Class-F_3 operation is also extended to mm-wave frequency generation in [32] and [33] which may interest a curious reader.

Table 3.3   Comparison with relevant oscillators

	This work	[9]	[8]	[29]	[10]	[30]	[2]	[20]
Technology	CMOS 65 nm	CMOS 130 nm	CMOS 350 μm	CMOS 65 nm	CMOS 55 nm	BiCMOS 0.130 μm	CMOS 90 nm	CMOS 65 nm
Supply voltage (V)	1.25	1	2.5	1.2	1.5	3.3	1.4	0.6
Frequency (GHz)	3.7[1]	5.2	1.2	3.92[1]	3.35[1]	1.56	0.915	3.7
Tuning range (%)	25	14	18	10.2	31.4	9.6	24.3	77
PN at 3 MHz (dBc/Hz)	−142.2	−141.2	−152	−141.7	−142	−150.4	−149	−137.1
Norm. PN[2] (dBc/Hz)	−154.3	−147.5	−154.8	−154.4	−153.3	−155	−149	−149.21
I_{DC} (mA)	12	1.4	3.74	18	12	88	18	17.5
Power consumption (mW)	15	1.4	9.25	25.2	27	290	25.2	10.5
FoM[3] (dB)	192.2	195	195	189.9	189	180	184.6	188.7
FoM$_T$[4] (dB)	200.2	198.4	200.7	190	199	179.7	192.3	206.5
Inductor/transformer count	1	1	2	2	1	1	1	1
Area (mm^2)	0.14	0.11	N/A	0.19	0.196	N/A	N/A	0.294
Oscillator structure	Class-F₃	Class-C	Noise filtering	Clip-and-restore	Class B/C	Colpitts	Traditional	Dual mode

[1] After on-chip ÷2 divider.
[2] Phase noise at 3-MHz offset frequency normalized to 915-MHz carrier.
[3] $FOM = |PN| + 20\log_{10}((f_0/\Delta f)) - 10\log_{10}(P_{DC}/1\text{mW})$.
[4] $FOM_T = |PN| + 20\log_{10}((f_0/\Delta f)(TR/10)) - 10\log_{10}(P_{DC}/1\text{mW})$.

3.5 Conclusion

We showed a LC-tank oscillator structure that introduces an impedance peak around the third harmonic of the oscillating waveform such that the third harmonic of the active device current converts into voltage and, together with the fundamental component, creates a pseudo-square oscillation voltage. The additional peak of the tank impedance is realized with a transformer-based resonator. As a result, the oscillator impulse sensitivity function reduces, thus lowering the conversion sensitivity of phase noise to various noise sources, whose mechanisms are analyzed in depth. Chief of these mechanisms arises when the active g_m-devices periodically enter the triode region during which the LC tank is heavily loaded while its equivalent quality factor is significantly reduced. The voltage gain, relative pole position, impedance magnitude, and equivalent quality factor of the transformer-based resonator are quantified at its two resonant frequencies. The gained insight reveals that the secondary to the primary voltage gain of the transformer can be even larger than its turns ratio. A comprehensive study of circuit-to-phase-noise conversion mechanisms of different oscillators' structures shows that the introduced class-F₃ exhibits the lowest phase noise at the same tank's quality factor and supply voltage. Based on this analysis, a class-F_3 oscillator was prototyped in a 65-nm CMOS technology. The measurement results proved expected performance of this oscillator in silicon.

References

[1] E. Hegazi and A. A. Abidii, "A 17-mW transmitter and frequency synthesizer for 900-MHz GSM fully integrated in 0.35-μm CMOS," *IEEE J. Solid-State Circuits*, vol. 38, no. 5, pp. 782–792, May 2003.

[2] R. B. Staszewski, J. L. Wallberg, S. Rezeq, C.-M. Hung, O. E. Eliezer, S. K. Vemulapalli, C. Fernando, K. Maggio, R. Staszewski, N. Barton, M.-C. Lee, P. Cruise, M. Entezari, K. Muhammad, and D. Leipold, "All-digital PLL and transmitter for mobile phones," *IEEE J. Solid-State Circuits*, vol. 40, no. 12, pp. 2469–2482, Dec. 2005.

[3] L. Vercesi, L. Fanori, F. D. Bernardinis, A. Liscidini, and R. Castello, "A dither-less all digital PLL for cellular transmitters," *IEEE J. Solid-State Circuits*, vol. 47, no. 8, pp. 1908–1920, Aug. 2012.

[4] H. Darabi, P. Chang, H. Jensen, A. Zolfaghari, P. Lettieri, J. C. Leete, B. Mohammadi, J. Chiu, Q. Li, S.-L. Chen, Z. Zhou, M. Vadipour, C. Chen, Y. Chang, A. Mirzaei, A. Yazdi, M. Nariman, A. Hadji-Abdolhamid, E. Chang, B. Zhao, K. Juan, P. Suri, C. Guan, L. Serrano, J. Leung, J. Shin, J. Kim, H. Tran, P. Kilcoyne, H. Vinh, E. Raith, M. Koscal, A. Hukkoo, V. R. C. Hayek, C. Wilcoxson, M. Rofougaran, and A. Rofougaran, "A quad-band GSM/GPRS/EDGE SoC in 65nm CMOS," *IEEE J. Solid-State Circuits*, vol. 46, no. 4, pp. 872–882, Apr. 2011.

[5] J. Borremans, G. Mandal, V. Giannini, B. Debaillie, M. Ingels, T. Sano, B. Verbruggen, and J. Craninckx, "A 40nm CMOS 0.4-6 GHz receiver resilient to out-of-band blockers," *IEEE J. Solid-State Circuits*, vol. 46, no. 7, pp. 1659–1671, Jul. 2011.

[6] J. Rael and A. Abidi, "Physical processes of phase noise in differential LC oscillators," *Proceedings of IEEE Custom Integrated Circuits Conference (CICC)*, 2000, pp. 569–572.

[7] P. Andreani, X. Wang, L. Vandi, and A. Fard, "A study of phase noise in Colpitts and LC-tank CMOS oscillators," *IEEE J. Solid-State Circuits*, vol. 40, no. 5, pp. 1107–1118, May 2005.

[8] E. Hegazi, H. Sjoland, and A. A. Abidi, "A filtering technique to lower LC oscillator phase noise," *IEEE J. Solid-State Circuits*, vol. 36, no. 12, pp. 1921–1930, Dec. 2001.

[9] A. Mazzanti and P. Andreani, "Class-C harmonic CMOS VCOs, with a general result on phase noise," *IEEE J. Solid-State Circuits*, vol. 43, no. 12, pp. 2716–2729, Dec. 2008.

[10] L. Fanori, A. Liscidini, and P. Andreani, "A 6.7-to-9.2 GHz 55 nm CMOS hybrid class-B/class-C cellular TX VCO," *IEEE International Solid-State Circuits Conference Digest of Technical Papers (ISSCC)*, Feb. 2012, pp. 354–355.

[11] H. Kim, S. Ryu, Y. Chung, J. Choi, and B. Kim, "A low phase-noise CMOS VCO with harmonic tuned LC tank," *IEEE Transactions on Microwave Theory and Techniques*, vol. 54, no. 7, pp. 2917–2923, Jul. 2006.

[12] M. Babaie and R. B. Staszewski, "Third-harmonic injection technique applied to a 5.87-to-7.56 GHz 65 nm class-F oscillator with 192 dBc/Hz FoM," *IEEE International Solid-State Circuits Conference Digest of Technical Papers (ISSCC)*, 2013, pp. 348–349.

[13] M. Babaie, and R. B. Staszewski, "A class-F CMOS oscillator," *IEEE J. Solid-State Circuits*, vol. 48, no. 12, pp. 3120–3133, Dec. 2013.

[14] A. Hajimiri and T. H. Lee, "A general theory of phase noise in electrical oscillators," *IEEE J. Solid-State Circuits*, vol. 33, no. 2, pp. 179–194, Feb. 1998.

[15] B. Razavi, "A millimeter-wave circuit technique," *IEEE J. Solid-State Circuits*, vol. 43, no. 9, pp. 2090–2098, Sept. 2008.

[16] J. R. Long, "Monolithic transformers for silicon RF IC design," *IEEE J. Solid-State Circuits*, vol. 35, no. 9, pp. 1368–1382, Sept. 2000.

[17] A. Bevilacqua, F. P. Pavan, C. Sandner, A. Gerosa, and A. Neviani, "Transformer-based dual-mode voltage-controlled oscillators," *IEEE Transactions on Circuits and Systems II, Exp. Briefs*, vol. 54, no. 4, pp. 293–297, Apr. 2007.

[18] A. Geol and H. Hashemi, "Frequency switching in dual-resonance oscillators," *IEEE J. Solid-State Circuits*, vol. 42, no. 3, pp. 571–582, Mar. 2007.

[19] B. Razavi, "Cognitive radio design challenges and techniques," *IEEE J. Solid-State Circuits*, vol. 45, no. 8, pp. 1542–1553, Aug. 2010.

[20] G. Li, L. Liu, Y. Tang, and E. Afshari, "A low-phase-noise wide-tuning-range oscillator based on resonant mode switching," *IEEE J. Solid-State Circuits*, vol. 47, no. 6, pp. 1295–1308, Jun. 2012.

[21] R. Degraeve, J. Ogier, R. Bellens, P. Roussel, G. Groeseneken, and H. Maesi, "A new model for the field dependence of intrinsic and

extrinsic time-dependent dielectric breakdown," *IEEE Transactions on Electron Devices*, vol. 45, no. 2, pp. 472–481, Feb. 2007.

[22] M. Babaie and R. Staszewski, "A Study of RF Oscillator Reliability in Nanoscale CMOS," *European Conference on Circuit Theory and Design (ECCTD)*, Sept. 2013 pp. 1–4.

[23] B. Razavi, "A study of phase noise in CMOS oscillators," *IEEE J. Solid-State Circuits*, vol. 31, no. 3, pp. 331–343, Mar. 1996.

[24] H. Krishnaswamy and H. Hashemi, "Inductor and transformer-based integrated RF oscillators: A comparative study," *Proceedings of IEEE Custom Integrated Circuits Conference (CICC)*, 2006, pp. 381–384.

[25] P. Andreani and J. R. Long, "Misconception regarding of transformer resonators in monolithic oscillator," *Electronic Letter*, vol. 42, no. 7, pp. 387–388, Mar. 2006.

[26] D. Murphy, J. J. Rael, and A. A. Abidi, "Phase noise in LC oscillators: A phasor-based analysis of a general result and of loaded Q," "Transformer-based dual-mode voltage-controlled oscillators," *IEEE Transactions on Circuits and Systems I, Reg. Papers*, vol. 57, no. 6, pp. 1187–1203, Jun. 2010.

[27] L. Fanori and P. Andreani, "Low-phase-noise 3.4–4.5 GHz dynamic bias class-C CMOS VCOs with a FoM of 191 dBc/Hz," *Proceedings of European Solid-state Circuits Conference (ESSCIRC)*, 2012, pp. 406–409.

[28] P. Andreani and A. Fard, "More on the phase noise performance of CMOS differentia lpair LC-tank oscillators," *IEEE J. Solid-State Circuits*, vol. 41, no. 12, pp. 2703–2712, Dec. 2006.

[29] A. Visweswaran, R. B. Staszewski, and J. R. Long, "A clip-and-restore technique for phase desensitization in a 1.2 V 65 nm CMOS oscillator for cellular mobile and base station," *IEEE International Solid-State Circuits Conference Digest of Technical Papers (ISSCC)*, 2012, pp. 350–351.

[30] J. Steinkamp, F. Henkel, P. Waldow, O. Pettersson, C. Hedenas, and B. Medin, "A Colpitts oscillator design for a GSM base station synthesizer," *Proceedings of IEEE Radio Frequency Integrated Circuits (RFIC) Symposium*, 2004, pp. 405–408.

[31] M. Babaie and R. B. Staszewski, "Class-F CMOS oscillator incorporating differntial passive network," *US Patent 9,197,221*, issued 24 Nov. 2015.

[32] Z. Zong, M. Babaie, and R. B. Staszewski, "A 60 GHz frequency generator based on a 20 GHz oscillator and an implicit multiplier," *IEEE*

Journal of Solid-State Circuits (JSSC), vol. 51, no. 5, pp. 1261–1273, May 2016.

[33] Z. Zong, P. Chen, and R. B. Staszewski, "A low-noise fractional-N digital frequency synthesizer with Implicit frequency tripling for mm-wave applications," *IEEE Journal of Solid-State Circuits (JSSC),* vol. 54, no. 3, pp. 755–767, Mar. 2019.

4

An Ultra-Low Phase Noise Class-F$_2$ CMOS Oscillator

In this chapter, we discuss and analyze a class of operation of an RF oscillator that further minimizes its phase noise. The main idea is to enforce a clipped voltage waveform around the LC tank by increasing the second harmonic of fundamental oscillation voltage through an additional impedance peak, thus giving rise to a class-F$_2$ operation. As a result, the noise contribution of the tail current transistor on the total phase noise can be significantly decreased without sacrificing the oscillator's voltage and current efficiencies. Furthermore, its special impulse sensitivity function (ISF) reduces the phase sensitivity to thermal circuit noise. The prototype of the class-F$_2$ oscillator is implemented in standard TSMC 65 nm CMOS occupying 0.2 mm^2. It draws 32–38 mA from 1.3-V supply. Its tuning range is 19% covering 7.2–8.8 GHz. It exhibits phase noise of -139 dBc/Hz at 3-MHz offset from 8.7-GHz carrier, translated to an average figure of merit of 191 dBc/Hz with less than 2-dB variation across the tuning range.

4.1 Introduction

Spectral purity of RF LC-tank oscillators is typically addressed by improving a quality factor (Q) of its tank, lowering its noise factor (NF) and increasing its power consumption. Even though technology scaling increases the *effective* capacitance ratio, $C_{max}/_{min}$, of switchable tuning capacitors and, consequently, the oscillator tuning range, it does not improve the oscillator's spectral purity parameters, such as tank Q-factor and oscillator NF. In fact, the tank Q-factor is slightly degraded in more advanced technologies mainly due to closer separation between the top metal and lossy substrate as well as

59

thinner lower-level metals that are used in metal-oxide-metal (MoM) capacitors. On the other hand, transistor noise factor keeps on degrading in more advanced technologies. Consequently, NF increases and thus penalizes the oscillator phase noise (PN). Consequently, the oscillators of excellent spectral purity and power efficiency are becoming more and more challenging as compared to other RF circuitry that is actually gaining from the technology scaling. This has motivated an intensive research leading to new oscillator topologies [1–11].

In this chapter, we specifically address the ultra-low phase noise design space while maintaining high power efficiency. We describe a soft-clipping class-F$_2$ oscillator topology based on enforcing a clipped voltage waveform around the LC tank by increasing the second-harmonic of the fundamental oscillation voltage through an additional impedance peak [8–10]. This structure shifts the oscillation voltage level so that it provides enough headroom for the tail current without compromising the oscillating amplitude. Consequently, the phase noise contribution of the tail current transistor is effectively reduced while maintaining the oscillator voltage efficiency. Furthermore, the class-F$_2$ operation clips the oscillation waveform for almost half of the period, thus benefiting from the lower circuit-to-phase-noise conversion during this time span.

The chapter is organized as follows: the trade-offs between the RF oscillator PN and power consumption are investigated in Section 4.2. Section 4.3 establishes the environment to introduce the class-F$_2$ operation, its benefits, and constraints. The circuit-to-PN conversion mechanisms are studied in Section 4.4. Section 4.5 presents extensive experimental results.

4.2 Challenges in Ultra-Low Phase Noise Oscillators

The phase noise (PN) of the traditional oscillator (i.e., class-B) with an ideal current source at an offset frequency $\Delta\omega$ from its fundamental frequency ω_0 could be expressed as

$$\mathcal{L}(\Delta\omega) = 10\log_{10}\left(\frac{KT}{2\,Q_t^2\,P_{DC}}\frac{1}{\alpha_I\,\alpha_V}\left(1+\gamma\right)\left(\frac{\omega_0}{\Delta\omega}\right)^2\right), \qquad (4.1)$$

where Q_t is the tank quality factor; α_I is the current efficiency, defined as the ratio of the fundamental current harmonic I_{ω_0} over the oscillator DC current I_{DC}; and α_V is the voltage efficiency, defined as the ratio of the oscillation amplitude V_{osc} (single-ended) over the supply voltage V_{DD}. The oscillator power consumption is

$$P_{DC} = \frac{V_{osc}^2}{\alpha_I \cdot \alpha_V \cdot R_{in}}, \tag{4.2}$$

where R_{in} is an equivalent input parallel resistance of the tank modeling its losses. Equation (4.1) clearly demonstrates a trade-off between power consumption and PN. To improve the oscillator PN, one must increase P_{DC} by scaling down R_{in}. This could be done by lowering the tank inductance while maintaining the optimal Q_t. For example, by keeping on reducing the inductance by half, R_{in} could theoretically decrease by half at the constant Q_t, which would improve phase noise by 3 dB with twice the power consumption at the same FoM.[1] However, at some point, the resistance of the tank's interconnections will start dominating the resonator losses and, consequently, the equivalent tank's Q will start decreasing. Hence, the PN-versus-power trade-off will no longer be beneficial since the FoM will drop dramatically due to the Q-factor degradation.

Coupling N oscillators is an alternative way of trading off the power for PN since it avoids scaling the inductance down to impractically small values. It can theoretically improve PN by a factor of N compared to a single oscillator [12, 13]. Unfortunately, the oscillator size increases linearly, i.e., $4\times$ larger area for just 6 dB of PN improvement.

In this chapter, we explain how to improve phase noise by utilizing two $1{:}n$ transformers that are connected back-to-back [8], as shown in Figure 4.1(b, c). The equivalent R_{in} and, thus, the oscillator PN are scaled down by a factor of $\sim(1 + n^2)$ without sacrificing tank's Q-factor. Hence, PN improvement can potentially be much better than with the coupled oscillators (e.g., Figure 4.1(a)) at the same die area. In addition, the C_2 and C_3 tuning capacitors, which are not directly connected to the primary of the first transformer, appear at the input of the transformer network via the scaling factor of n^2 and n^4 as can be realized from Figure 4.1(c). This impedance transformation results in a significant reduction in the required value of all the capacitors (i.e., $\sum_i C_i$), which reduces the routing parasitics (both inductive and capacitive), and improves the tuning range and PN of the oscillator. Even though by increasing the transformer's turns ratio the tank input impedance will be reduced, the transformer Q-factor will not stay at the optimum level and will start dropping at some point [14]. It turns out that the turns ratio of $n = 2$ can satisfy the aforementioned constraints altogether.

---

[1]FoM $= |\text{PN}| + 20\log_{10}(\omega_0/\Delta\omega) - 10\log_{10}(P_{DC}/1\text{mW})$.

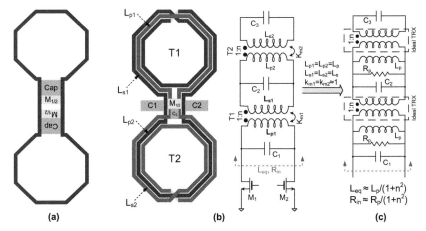

Figure 4.1 Phase noise reduction techniques without sacrificing tank's Q-factor: (a) coupled oscillators, (b) connecting two step-up transformers back-to-back, and (c) its equivalent circuit model.

To sustain the oscillation of this differential tank, two transistors shall be added. Figure 4.2 illustrates the *preliminary* schematic and waveforms. Unfortunately, as gathered from Figure 4.3, this structure suffers the same issues as the traditional class-B oscillator when the ideal current source is replaced with a tail bias transistor, M_T. The PN is ideally improved by 20 dB/dec through increasing the oscillation amplitude, provided the gm-devices $M_{1,2}$ operate in saturation over the entire period. However, the slope of PN improvement deviates from the ideal case when $M_{1,2}$ enter the triode region for a part of the oscillation period [15]. This problem is intensified especially when the oscillator operates at higher frequencies and larger I_{DC} (i.e., ≥ 10 mA) is needed to satisfy the stringent spectral purity of the GSM standard [16]. Actually, the combination of the parasitic drain capacitance of the large-size M_T with the entering the triode region by $M_{1,2}$, cyclically short-circuits the tank, thus degrading its equivalent Q-factor and oscillator PN [17].

Furthermore, the oscillation voltage should provide minimum V_{DSAT} across M_T throughout the entire period to keep it in saturation. Consequently, α_V becomes substantially less than 1, which translates to a significant PN penalty as clearly seen from (4.1). Larger M_T needs lower V_{DSAT}, which would increase α_V. However, the tail transistor's effective thermal noise will increase significantly for the same I_{DC} [18]. As a consequence, the contribution of M_T to the PN could be larger than that of gm-devices, which

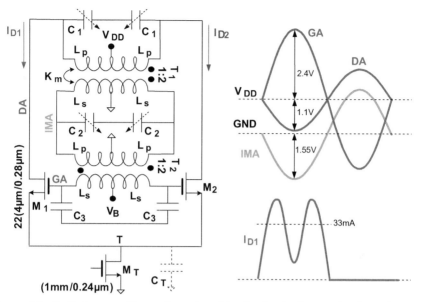

Figure 4.2 Preliminary oscillator schematic and its simulated voltage and estimated current waveforms at $f_0 = 8$ GHz, $V_{DD} = 1.2$ V, $I_{DC} = 33$ mA, $L_{eq} = 80$ pH, and $C_{eq} = 4.95$ pF.

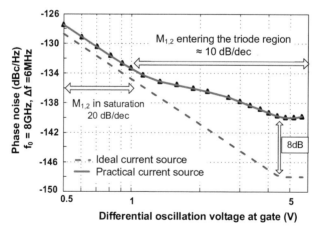

Figure 4.3 Simulated phase noise performance of the preliminary oscillator of Figure 4.2 versus gate differential oscillation voltage for the ideal and real current sources.

translates to a significant increase of the oscillator NF and thus its PN [16]. In addition, the M_T parasitic capacitance, C_T, will also increase with the side effect of a stronger tank loading. On the other hand, the combination of the sinusoidal drain voltage, large C_T, and the entering of triode region by $M_{1,2}$ will result in a dimple in the squarish shape of active device drain current (see Figure 4.2) with a 10%–20% reduction in α_I and thus FoM of the oscillator [1, 16]. All the above reasons contribute to reducing the rate of PN improvement versus V_{osc} to about 10 dB/dec when $M_{1,2}$ enter the triode region for a part of the oscillation period. Hence, a huge 8-dB PN difference in Figure 4.3 is observed between the ideal and real operation of the oscillator. Consequently, the novel oscillator must not be sensitive to the excess gm-device noise in the triode intervals. It should also break the trade-off between α_V and NF.

4.3 Evolution Towards Class-F₂ Operation

Before introducing a new phase noise (PN) reduction technique, let us take a closer look at the harmonic component of the drain current I_D of the M_1 and M_2 gm-devices in Figure 4.2. Ideally, I_D is a square wave containing fundamental and odd harmonics. The odd harmonics through M_1 and M_2 are 180° mutually out-of-phase and appear as differential-mode (DM) input signals for the tank. The I_D also contains even harmonics due to the large oscillation voltage, non linearity of $M_{1,2}$, and large parasitic capacitance of M_T. However, the even harmonics through M_1 and M_2 are mutually in-phase with ±90° phase shift to their related odd harmonics, as shown in Figure 4.4. Consequently, these even harmonics appear as a common-mode (CM) input for the tank. The conventional tank input impedance has only one

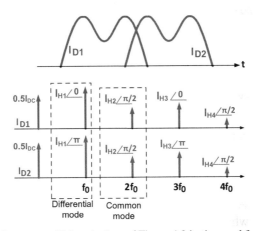

Figure 4.4 Drain current of $M_{1,2}$ devices of Figure 4.2 in time and frequency domains.

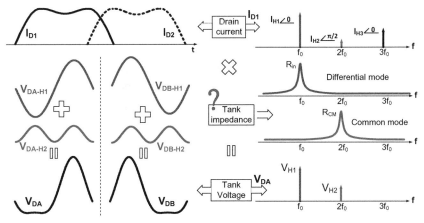

Figure 4.5 New oscillator's waveforms in time and frequency domains.

peak at the fundamental frequency ω_0. Therefore, the tank filters out the drain current harmonics and ultimately a sinusoidal voltage is seen across the tank. Now, suppose the tank offers an additional CM input impedance peak around the second harmonic (see Figure 4.5). Then, the second harmonic of I_D is multiplied by the tank's CM input impedance to produce a sinusoidal voltage at $2\omega_0$ that is in quadrature to the fundamental oscillation voltage produced by the tank's DM impedance at ω_0. The combination of both waveforms creates the desired oscillation voltage around the tank, as shown in Figure 4.5, thus justifying the class-F₂ designation.

$$V_{DA} = V_{DD} - V_{H1} \sin(\omega_0 t) - V_{H2} \sin\left(2\omega_0 t + \frac{\pi}{2}\right) \qquad (4.3)$$

ζ_V is defined as the ratio of the second-to-first harmonic components of the oscillation voltage.

$$\zeta_V = \frac{V_{H2}}{V_{H1}} = \left(\frac{R_{CM}}{R_{in}}\right)\left(\frac{I_{H2}}{I_{H1}}\right), \qquad (4.4)$$

where R_{in} and R_{CM} are, respectively, the tank DM and CM impedance magnitude at ω_0 and $2\omega_0$. Figure 4.6 illustrates the oscillation voltage and its related impulse sensitivity function (ISF) based on Equation (38) in [21] for different ζ_V values. Clearly, ζ_V should be 0.3 to have the widest flat span in the tank's oscillation voltage. The Γ^2_{rms} is 0.35 for $\zeta_V = 0.3$ compared to 0.5 for the traditional oscillator, which leads to a 1.5-dB PN and FoM improvements. Furthermore, ISF is negligible when the gm-devices work in the triode region

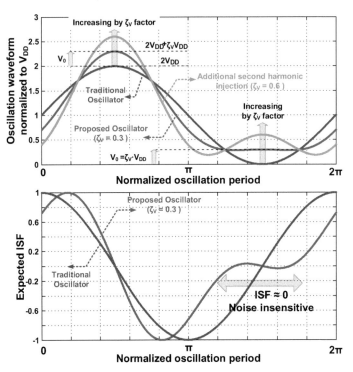

Figure 4.6   Effect of adding second harmonic in the oscillation voltage waveform (top) and its expected ISF based on Equation (38) in [21] (bottom).

and inject the most thermal noise into the tank. Consequently, the oscillator FoM improvement should be larger than that predicted by just the ISF$_{rms}$ reduction. More benefits of the class-F$_2$ operation will be revealed in the following sections.

The argument related to Figure 4.5 suggests the creation of an additional CM input impedance peak at the second harmonic of main differential resonance. Incidentally, the introduced step-up 1:2 transformer acts differently to the CM and DM input signals. Figure 4.7(a) illustrates the induced current at the transformer's secondary when the primary winding is excited by a differential signal. All induced currents circulate in the same direction at the transformer's secondary to satisfy Lenz's Law. Consequently, the induced currents add constructively, which leads to a strong inter-winding coupling factor (k$_m$ ≥ 0.7). However, when the transformer's primary is excited by a CM signal (Figure 4.7(b)), the induced currents at the right-hand and left-hand halves of the transformer's secondary winding circulate in opposite directions, thus

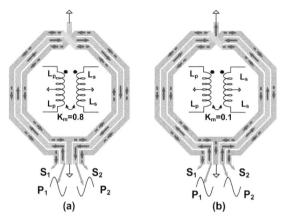

Figure 4.7 Transformer behavior in (a) differential-mode and (b) common-mode excitations.

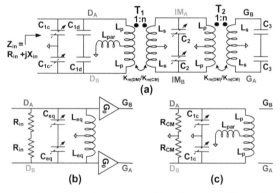

Figure 4.8 New transformer-based resonator: (a) schematic, (b) its simplified equivalent differential-mode circuit ($k_{m(DM)} \approx 1$), and (c) simplified tank schematic for common-mode input signals ($k_{m(CM)} \approx 0$).

largely canceling each other. The residual current results in a very small $k_m \leq 0.2$ for the CM excitation. Consequently, the concept of using two modes of a transformer for waveform shaping (proposed in [19] for a power amplifier) will be adopted here to realize the special tank input impedance of Figure 4.5. Note that an equivalent lumped-element model in [14, 20] cannot simultaneously cover both CM and DM types of behavior, and would produce wrong results. Hence, we suggest to utilize the transformer's S-parameters and PSS analysis to simulate the novel class-F$_2$ oscillator.

Figure 4.8 shows the newly invented tank of a class-F$_2$ oscillator. The C_{1d} and C_3 are intentionally chosen as fixed capacitors while the DM and

CM resonant frequencies are tuned by C_{1c} (fine) and C_2 (coarse). The DM main resonant frequency is

$$f_{1d} = \frac{1}{2\pi\sqrt{L_{eq}C_{eq}}} \approx \frac{1}{2\pi\sqrt{\left(\frac{L_p}{1+n^2}\right)(C_{1c} + C_{1d} + C_2 n^2 + C_3 n^4)}}. \quad (4.5)$$

The inductance reduction and capacitance multiplication factors of the dual-transformer tank are directly contained in (4.5). The CM input signal can neither see the second transformer nor C_2 and C_3 due to negligible $k_{m(CM)}$. In addition, differential capacitors also act as open circuit for the CM signals. Consequently, the tank's CM resonant frequency is

$$f_{1c} = \frac{1}{2\pi\sqrt{L_{cm}C_{1c}}} \approx \frac{1}{2\pi\sqrt{(L_p + 2L_{par})C_{1c}}}. \quad (4.6)$$

There is no tank impedance scaling for the CM excitation. Hence, the CM input impedance peak should be higher than the DM peak, as clearly seen from Figure 4.9 (top). To operate properly, CM-to-DM resonance ratio must be adjusted to 2:

$$\zeta_f = \frac{f_{1c}}{f_{1d}} = \sqrt{\frac{L_p}{L_p + 2L_{par}} \cdot \frac{C_{1c} + C_{1d} + C_2 n^2 + C_3 n^4}{C_{1c}(1 + n^2)}} = 2. \quad (4.7)$$

As a consequence, the frequency tuning requires a bit different consideration than in the class-B oscillators. Both C_{1c} and C_2 must, at least at the coarse level, be changed simultaneously to satisfy (4.7) such that f_{1c} coincides with $2f_{1d}$. This adjustment is entirely a function of the ratio of the tuning capacitors, which is precise, thus making ζ_f largely independent from process, voltage, and temperature (PVT) variations.

Let us now consider the required accuracy of this ratio ζ_f. The transformer and switching capacitors are designed based on maximum Q-factor at the operating frequency f_{1d}. The tank Q-factor drops at least $3\times$ at $f_{1c} = 2f_{1d}$. Consequently, the tank CM impedance bandwidth is very wide, as seen in Figure 4.9. Therefore, the oscillator is less sensitive to the position of f_{1c} and thus the tuning capacitance ratio. A realistic 5% error in ζ_f has no significant adverse effects on the oscillator waveform and thus its PN.

The schematic and waveforms of the new oscillator are shown in Figures 4.10 and 4.11. Even though the second harmonic injection reduces the drain oscillation voltage by V_0 during the negative clipping interval,

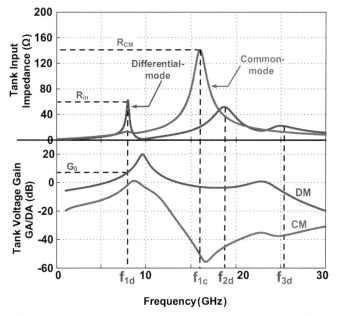

Figure 4.9 Simulated characteristics of the transformer-based tank of Figure 4.8: (top) magnitude of input impedance Z_{in}; (bottom) tank voltage gain between gate and drain of core devices.

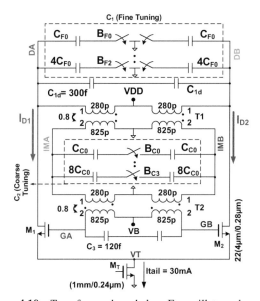

Figure 4.10 Transformer-based class-F_2 oscillator schematic.

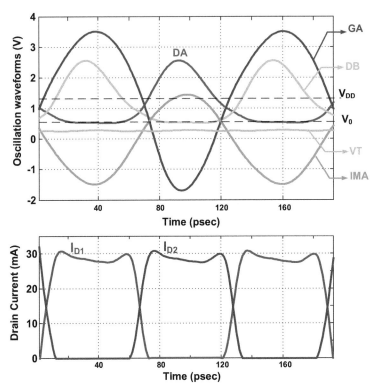

Figure 4.11   Simulated oscillation waveforms of the class-F_2 oscillator at $V_{DD} = 1.2$ V and $I_{DC} = 29$ mA: (top) oscillation voltage of different circuit nodes and (bottom) core transistors drain current.

it increases its positive peak by V_0 (see Figure 4.6(a)). It means the drain oscillation span is shifted from 0-to-$2V_{DD}$ in the traditional oscillator to V_0-to-$(2V_{DD} + V_0)$ in the class-F_2 operation. Hence, the larger current source voltage headroom and lower noise factor are achieved without compromising the oscillation amplitude. Furthermore, the V_0 headroom also reduces the dimple in the core-device drain current (compare Figures 4.2 and 4.11), which helps the class-F_2 current efficiency to be closer to the ideal value of $2/\pi$.

Figure 4.9 illustrates the tank CM/DM input impedance and passive voltage gain between the gate and drain of $M_{1,2}$ versus frequency. Unfortunately, the tank exhibits two other undesired DM resonant frequencies (f_{2d}, f_{3d}) due to imperfect k_m of the two transformers that create two leakage inductances [14]. Consequently, the circuit loop must guarantee the

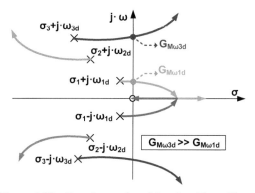

Figure 4.12 Root-locus plot of the class-F$_2$ oscillator.

oscillation only at the desired DM resonance, f$_{1d}$. Although CM demonstrates much larger input impedance peak, the two transformers effectively reject (attenuate by >40 dB) the CM signals. The root-locus plot in Figure 4.12 illustrates the DM pole movements toward zeros for different oscillator loop transconductance gains G$_M$. The first and third frequency conjugate pole pairs (ω_{1d}, ω_{3d}) move into the right-hand plane by increasing the absolute value of G$_M$, while the second conjugate pole ω_{2d} is pushed far away from the imaginary axis. This guarantees that the oscillation will not happen at ω_{2d}. Furthermore, it can be shown that ω_{3d} poles move to much higher frequencies with much lower input impedance peak and tank voltage gain if enough differential capacitance is located at T$_1$ primary windings. It justifies the existence of the non-switchable differential capacitor C$_{1d}$. Consequently, the loop gain will not be enough to satisfy the Barkhausen criterion for ω_{3d}.

4.4 Phase Noise Mechanism in Class-F$_2$ Oscillator

According to the linear time-variant (LTV) model [21], the phase noise of an oscillator at an offset frequency $\Delta\omega$ from its fundamental frequency ω_0 is expressed as

$$\mathcal{L}(\Delta\omega) = 10\log_{10}\left(\frac{\sum_i N_{L,i}}{2\,q_{max}^2\,(\Delta\omega)^2}\right), \tag{4.8}$$

where $q_{max} = C_{eq}\cdot V_{osc}$ is the maximum charge displacement across the tuning capacitors and $V_{osc} = \alpha_I\cdot R_{in}\cdot I_{DC}$ is the single-ended oscillation amplitude at the drain of gm-devices. The $N_{L,i}$ in (4.8) is the effective noise power produced by ith device given by

$$N_{L,i} = \frac{1}{2\pi} \int_0^{2\pi} \Gamma_i^2(\omega_0 t) \, \overline{i_{n,i}^2(\omega_0 t)} d(\omega_0 t), \qquad (4.9)$$

where $\overline{i_{n,i}^2}$ is the white noise current density of the ith noise source and Γ_i is its corresponding ISF function. Obtaining the ISF of various oscillator nodes is the first step in calculating the oscillator's PN. The ISF functions are simulated by injecting a 20 femto-coulomb charge (Δq) throughout the oscillation period and measuring the resulting time shifts, Δt_i.

$$\Gamma_i = \omega_0 \cdot \Delta t_i \cdot \frac{q_{max}}{\Delta q} \qquad (4.10)$$

Figure 4.13(a) illustrates the ISF of various tank nodes. The soft clipping reduces by 30% the effect of losses on the oscillator PN due to single-ended switchable C_{1c}[2] and T_1 primary windings. However, ISF functions of the T_1 secondary and T_2 primary/secondary winding noise sources (including C_2 and C_3) are not improved due to the sinusoidal (i.e., conventional) waveforms at $IM_{A,B}$ and $G_{A,B}$ nodes. Figure 4.13(a) indicates that $G_{A,B}$ are the most sensitive nodes. Hence, C_3 is constructed as a fixed MoM capacitor and the transformer was designed with a goal of maximizing Q-factor of the secondary winding.

To calculate a closed-form PN equation, the oscillator model is simplified in Figure 4.14. The $G_{DS1,2}(t)$ represent the channel conductance of $M_{1,2}$. The $G_{M1,2}(t)$ and $G_{MT}(t)$ model the transconductance gain of $M_{1,2}$ and M_T, respectively. The original tank is pruned to a parallel L_{eq}, C_{eq}, R_{in} with noise-less voltage gain of G_0 (see Figure 4.8(b)). The simplified tank's equivalent ISF can be roughly estimated by an average of the tank's contributing ISF functions of Figure 4.13(a) and is shown in Figure 4.13(b) as green curve. The effective noise power of the tank is illustrated in Figure 4.13(c) as green curve and its average power is approximated by

$$N_{Tank} = \frac{1}{\pi} \int_0^{2\pi} \frac{4KT}{R_{in}} \Gamma_{tank}^2(\omega_0 t) d(\omega_0 t) \approx 0.8 \frac{KT}{R_{in}}. \qquad (4.11)$$

Consequently, the soft clipping reduces N_{Tank} by 20% compared to the traditional oscillator.

---

[2]The single-ended switchable capacitor is used to adjust the CM resonant frequency. However, its Q-factor is almost half of that of the differential structure for the same C_{max}/C_{min}. The soft clipping largely compensates the effect of additional losses due to its lower Γ_{rms} value.

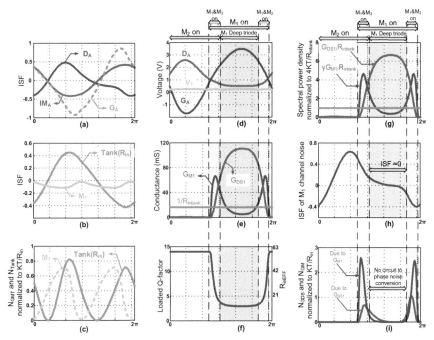

Figure 4.13 Mechanisms of circuit-to-phase-noise conversion across the oscillation period in the class-F_2 oscillator: (a) simulated ISF of different tank nodes, (b) equivalent ISF in the simplified oscillator schematic of Figure 4.14, (c) simulated effective power spectral density of the oscillator's noise sources normalized to KT/R_{in}, (d) oscillation waveforms and operation region of $M_{1,2}$, (e) transconductance and channel conductance of M_1, (f) loaded Q-factor and effective parallel input resistance of the tank, (g) power spectral density of M_1 noise sources normalized to $4KT/R_{in}$, (h) simulated ISF function of M_1 channel noise, and (i) simulated effective power spectral density of different noise sources of M_1 normalized to KT/R_{in}.

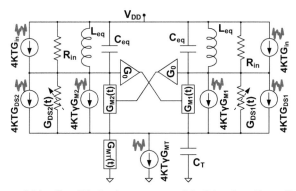

Figure 4.14 Simplified noise source model of the class-F_2 oscillator.

The effects of noise on the oscillator PN due to channel conductance (G_{DS}) and transconductance gain (G_M) of $M_{1,2}$ transistors are separately investigated. Figure 4.13(d) illustrates various operational regions of $M_{1,2}$ across the oscillation period. When $M_{1,2}$ are not turned off, they work mainly in the deep triode region where they exhibit a few ohms of channel resistance, as indicated in Figure 4.13(e). Consequently, the combination of the large parasitic capacitance of M_T with low channel resistance of $M_{1,2}$ in this deep triode region makes a low impedance path between the tank and ground. The literature interprets this as the tank loading event and defines implicit parameters such as *effective* tank Q-factor (Q_{eff}) and input parallel resistance (R_{ineff}) to justify the oscillator phase noise degradation due to this phenomenon. As shown in Figure 4.13(f), the tank Q_{eff} and R_{ineff} drop 4–5× when $M_{1,2}$ operate in the deep triode. These "effective" parameters merely indicate that more noise is injected then into the tank. However, they do not ordain how much of that circuit noise converts to phase noise, especially when the drain oscillation wave is not conventionally sinusoidal.

The proper approach should be based on the channel conductance noise power and its related ISF. If we had an ideal current source, $M_{1,2}$ noise would be injected to the tank only during the commutating time (hachure areas in Figure 4.13(e–g)). At the remaining part of oscillation period, one transistor is off and the other one is degenerated by the ideal current source and thus noiseless. However, the output impedance of a practical current source is low for such a high $I_{DC} = 30$ mA and $f_0 = 8$ GHz. Consequently, $M_{1,2}$ can inject significant amount of noise to the tank when they operate in deep triode region outside the commutating time (i.e., gray area in Figure 4.13(g)). Note that gm-devices generate ∼7× higher amount of noise compared to the tank loss in the gray area, which can potentially increase the phase noise of the oscillator. However, the ISF of channel noise of $M_{1,2}$ is very small in that time span as shown in Figure 4.13(h). Hence, the excessive transistor channel noise (or excessive loaded tank noise of the conventional approach) cannot convert to phase noise. Consequently, the effective noise power of the gm-device channel conductance is negligible, as illustrated in Figure 4.13(i), and its average power is approximated by

$$N_{GDS} = \frac{1}{\pi} \int_0^{2\pi} 4KTG_{DS1}(\omega_0 t)\cdot\Gamma_{M1}^2(\omega_0 t)\cdot d(\omega_0 t) \approx \frac{KT}{R_{in}}\cdot(0.25) \quad (4.12)$$

Note that N_{GDS} is at least 4× larger for the traditional oscillator, especially when a large α_V is needed [22].

Figure 4.13(e) shows M_1 transconductance gain G_{M1} across the oscillation period. To sustain the oscillation, the combination of the transformers' passive voltage gain (G_0) and effective negative transconductance of the gm-devices needs to overcome the tank and $M_{1,2}$ channel resistance losses. Consequently,

$$G_{M1EF} = \frac{1}{G_0} \cdot \left(\frac{1}{R_{in}} + G_{DS1EF} \right), \qquad (4.13)$$

where G_{DS1EF} describes the effective value of the instantaneous conductance $G_{DS1}(t)$ of $M_{1,2}$ [22]. It can be shown that G_{DS1EF} could be as large as $1/R_{in}$ when the oscillator is biased near the voltage limited region [1]. Therefore, the effective noise due to G_M of core transistors can be calculated by

$$N_{GM} = \frac{1}{\pi} \int_0^{2\pi} 4KT\gamma_{M1} G_{M1}(\omega_0 t) \cdot \Gamma_{M1}^2(\omega_0 t) \cdot d(\omega_0 t)$$

$$\approx \frac{KT}{R_{in}} \frac{\gamma_{M1}}{G_0} \cdot (1 + R_{in} \cdot G_{DS1EF}) \approx \frac{KT}{R_{in}} \cdot \left(\frac{2\gamma_{M1}}{G_0} \right) \qquad (4.14)$$

Equation (4.14) indicates that the second harmonic injection (i.e., class-F$_2$ operation) demonstrates no benefit for N_{GM}, but the transformers' voltage gain still offers significant benefits.

To estimate the PN contribution of M_T, its transconductance should be calculated first.

$$G_{MT} = \frac{2I_{DC}}{V_{gs(M_T)} - V_{th}} \approx \frac{2I_{DC}}{V_T}, \qquad (4.15)$$

where V_T is the overdrive voltage of M_T equal to the drain–source voltage. The clipping voltage level is

$$V_0 = V_{DD} \left[1 - \alpha_V \left(1 - \zeta_V \right) \right]. \qquad (4.16)$$

By dedicating a half of V_0 headroom to M_T,

$$G_{MT} = \frac{4I_{DC}}{V_0} \approx \frac{4I_{DC}}{V_{DD} \left(1 - \alpha_V \left(1 - \zeta_V \right) \right)}. \qquad (4.17)$$

By substituting I_{DC} with $V_{osc}/\left(\alpha_I R_{in} \right)$ in (4.17),

$$G_{MT} = \frac{4}{\left(1 - \alpha_V \left(1 - \zeta_V \right) \right) R_{in}\alpha_I} \left(\frac{V_{osc}}{V_{DD}} \right) = \frac{1}{R_{in}} \frac{4\alpha_V}{\left(1 - \alpha_V \left(1 - \zeta_V \right) \right) \alpha_I}. \qquad (4.18)$$

As discussed earlier, α_I and α_V could be as large as 0.6 and 0.9 in this oscillator, and optimum ζ_V is about 0.3. Hence, (4.18) is simplified to $G_{MT} \approx 15/R_{in}$. As revealed by Figure 4.13(b, orange), $\Gamma_{MT,rms}$ is only 0.08 due to relatively large V_T of the class-F$_2$ operation. Consequently,

$$N_{MT} = \frac{1}{2\pi} \int_0^{2\pi} 4KT\gamma_{MT}G_{MT}(\omega_0 t)\cdot\Gamma_{MT}^2(\omega_0 t)\cdot d(\omega_0 t) \approx \frac{KT}{R_{in}}\cdot(0.4\gamma_{MT}).$$

(4.19)

The contribution of M_T to the PN is less than that of the tank and is about 20% of the total. This share could easily be higher than 50% for the traditional oscillator at the same α_V and I_{DC} as discussed in [16, 17]. Finally, the total oscillator effective noise power (N_T) and noise factor (NF_{Total}) are given by

$$N_T \approx \frac{KT}{R_{in}}\cdot NF_{Total}, \qquad NF_{Total} \approx \left(1.05 + \frac{2\gamma_{M1}}{G_0} + 0.4\gamma_{MT}\right).\quad(4.20)$$

Equation (4.20) indicates that the effective noise factor of the class-F$_2$ oscillator is very close to the ideal value of $(1+\gamma)$ despite the aforementioned practical issues. The phase noise can easily be calculated by replacing (4.20) in (4.8). The oscillator FoM normalizes the PN performance to ω_0 and P_{DC}, yielding

$$FoM = -10\log_{10}\left(\frac{10^3 \cdot KT}{2Q_t^2\alpha_I\alpha_V}\cdot NF_{Total}\right).\quad(4.21)$$

Table 4.1 verifies the solidity of the presented phase noise analysis by comparing the results of SpectreRF^{TM} PSS, Pnoise simulations with the derived theoretical equations. The expressions estimate the oscillator PN and share of different noise sources with an acceptable accuracy.

It is also instructive to compare in Table 4.2 the benefits and drawbacks of the two flavors of class-F operation. Intuitively, the third-harmonic injection in class-F$_3$ (Chapter 3) demonstrates a pseudo-square waveform with smaller ISF$_{rms}$ value and shorter commutating time. Consequently, it offers lower NF$_{Tank}$ and NF$_{GM}$. On the other hand, class-F$_2$ operation provides larger voltage overhead for the gm-devices and tail current transistor without sacrificing the oscillator α_V. Hence, it exhibits better NF$_{MT}$, NF$_{GDS}$, and α_V. As expected, the effective noise factor and FoM of both topologies turns out to be identical. However, this implementation of class-F$_2$ automatically scales down the tank input parallel resistance and thus offers lower PN at price of larger area and slightly lower Q-factor due to the interconnection of the two transformers.

Table 4.1 Comparison between the results of SpectreRF PSS, Pnoise simulation and theoretical equations at 8-GHz carrier for $V_{DD} = 1.2$ V, $R_{in} = 60\ \Omega$, $L_{eq} = 80$ pH, $\gamma_{MT} = 1.3$, and $\gamma_{M1,2} = 1$

	Theoretical Equations		SpectreRF Simulation	
	Value	Share	Value	Share
N_{Tank}	$5.50 \cdot 10^{-23}$ V^2/Hz	31%	$4.71 \cdot 10^{-23}$ V^2/Hz	28.4%
$N_{M1,2} =$ $N_{GDS} + N_{GMT}$	$8.63 \cdot 10^{-23}$ V^2/Hz	48.8%	$8.78 \cdot 10^{-23}$ V^2/Hz	53%
N_{MT}	$3.59 \cdot 10^{-23}$ V^2/Hz	20.2%	$3.08 \cdot 10^{-23}$ V^2/Hz	18.6%
$N_T =$ $N_{Tank} + N_{M1,2} + N_{MT}$	$17.72 \cdot 10^{-23}$ V^2/Hz	100%	$16.57 \cdot 10^{-23}$ V^2/Hz	100%
q_{max} (coulumbs)	$5.34 \cdot 10^{-12}$		$5.34 \cdot 10^{-12}$	
Phase noise @10MHz	-151 dBc/Hz		-151.33 dBc/Hz	

Table 4.2 Comparison between two flavors of class-F oscillator for the same carrier frequency = 8 GHz, $V_{DD} = 1.2$ V, tank Q-factor = 14, $\Delta f = 10$ MHz, and $R_P = 240\ \Omega$

	Expression	Class-F$_3$	Class-F$_2$
α_I	I_{ω_0}/I_{DC}	0.6	0.6
α_V	V_{osc}/V_{DD}	0.8	0.9 (✓)
NF_{Tank}	$N_{Tank}/(KT/R_{in})$	0.7 (✓)	0.8
NF_{GDS}	$N_{GDS}/(KT/R_{in})$	0.3	0.25 (✓)
NF_{GM}	$N_{GM}/(KT/R_{in})$	$\approx 0.7\gamma_{M1}$ (✓)	$\approx \gamma_{M1}$
NF_{GMT}	$N_{GMT}/(KT/R_{in})$	$\approx 0.5\gamma_{MT}$	$\approx 0.4\gamma_{MT}$ (✓)
NF_{Total}		3.7dB (✓)	4.1dB
FoM	$\approx -10\log_{10}\left(\frac{KT}{2Q_t^2\,\alpha_I\,\alpha_V}NF\right)$	192.9dB (✓)	192.9dB (✓)
R_{in}		$\approx R_p = 240\Omega$	$\approx R_p/\left(1+n^2\right) = 60\Omega$ (✓)
P_{DC}	$\left(\frac{V_{DD}^2}{R_{in}}\frac{\alpha_V}{\alpha_I}\right)$	8mW	36mW
Phase noise	$\approx 10\log_{10}\left(\frac{KT}{2Q_t^2\,P_{DC}}\frac{1}{\alpha_I\,\alpha_V}NF\left(\frac{\omega_0}{\Delta\omega}\right)^2\right)$	-144 dBc/Hz	-150.5 dBc/Hz (✓)

4.5 Experimental Results

This oscillator targets GSM-900 MHz and DSC-1800 MHz base-station PN requirements. Electromagnetic (EM) simulations reveal that the tank Q-factor would be slightly (i.e., ~10%) better at 8 GHz as compared to 4 GHz for the

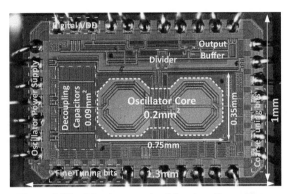

Figure 4.15   Die photograph of the class-F$_2$ oscillator.

same R_{in} and tuning range. However, the 1/f noise upconversion would be more severe at 8 GHz due to a larger share of the non linear C_{gs} of gm-devices to the total tank's capacitance. Furthermore, the output impedance of the current source is lower at higher frequencies, which would lead to higher PN penalty due to the tank loading. Consequently, there seems to be altogether no clear performance advantage of the 8 GHz over 4 GHz operation. Considering the fact that this oscillator has two transformers, the 8 GHz center frequency was chosen mainly to save die area.

The class-F$_2$ oscillator, whose schematic was shown in Figure 4.10, was realized in TSMC 1P7M 65 nm CMOS process technology. The die photograph is shown in Figure 4.15. The oscillator core die area is 0.2 mm^2. The differential transistors are thick-oxide devices of 22(4 μm/0.28 μm) dimension. However, the tail current source M_T is implemented as a standard 1 mm/0.24 μm thin-oxide (t_{ox} = 2.6 nm) device. Note that the thin-oxide device produces lower 1/f noise corner than the thick one at the same area [23]. The aluminum capping layer (1.45 μm), which is intended to cover bond pads, is strapped to the ultra-thick top copper layer (3.4 μm) to form the windings and improve the transformer's primary and secondary Q-factor to 14 and 20, respectively, at 8 GHz. The transformer's primary and secondary differential self-inductance is 560 and 1650 pH, respectively, with the DM and CM magnetic coupling factors of 0.8 and 0.15, respectively.

Four differential switched MOM capacitors $B_{C0}-B_{C3}$ with the resolution of 40 fF placed across T_1 secondary realize coarse tuning bits (C_2), while the fine control bits $B_{F0}-B_{F2}$ with LSB of 20 fF adjust ω_c to near $2\omega_{1d}$ to satisfy (4.7) and thus the class-F$_2$ operation. The effective C_{max}/C_{min} of the switched capacitor structures is determined by the strong trade-off between the oscillator tuning range (TR) and tank Q-factor degradation due to the switch series resistance. The switched-capacitor's Q-factor is about 45 for 25% TR at 8 GHz. Furthermore, the interconnections of the two transformers also increase the tank losses by 10%, resulting in an average Q-factor of 14 for the entire tank.

The supply voltage connects to the center tap of T_1 primary along with a 100 pF on-chip decoupling capacitor. The center tap of T_2 secondary is connected to the bias voltage V_B, which is fixed at V_{DD} to minimize the number of supply domains and to guarantee safe oscillator start-up. The oscillator is very sensitive to noise at the $M_{1,2}$ gates (see Figure 4.13(a)). Fortunately, no DC current is drawn from V_B, so an RC filter of slow time constant is placed between V_{DD} and V_B to further reduce the bias voltage noise. Both T_1 secondary and T_2 primary winding center taps are connected to ground to avoid any floating nodes and make a return path for the negligible second-harmonic current to improve the waveform symmetry.

The measured and simulated PN at 4.35 GHz (after the on-chip ÷2 divider) at 1.3 V and 32 mA current consumption are shown in Figure 4.16.

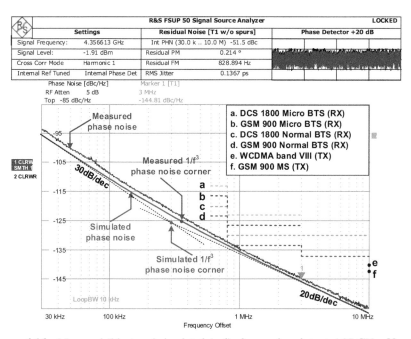

Figure 4.16 Measured (blue) and simulated (red) phase noise plots at 4.35 GHz, $V_{DD} =$ 1.3 V and $P_{DC} = 41$ mW. Specifications (MS: mobile station, BTS: basestation) are normalized to the carrier frequency.

The PN of −145 dBc/Hz at 3 MHz offset lies on the 20 dB/dec region, which extrapolates to −174.7 dBc/Hz at 20 MHz offset (normalized to 915 MHz) and meets the GSM TX mobile station (MS) requirements with a very wide 13 dB margin. The GSM/DCS "micro" base-station (BTS) and DCS "normal" BTS specs are met with a few dB of margin. These PN numbers are the *best ever* published at low V_{DD} (i.e., ≤1.5 V). However, the toughest GSM base-station "normal" specifications at 800-to-900 kHz offset are within 1 dB of reach. The measured PN is just 1 dB higher than simulation in the 20-dB/dec region due to the power supply noise and additional tank loss caused by the routing of the tuning capacitors and dummy fill metals around the transformer.

The measured $1/f^3$ PN corner shows less than 100 kHz increase over the simulation and is ∼350 and ∼250 kHz at the highest and lowest side of the tuning range, respectively. This excellent $1/f^3$ performance is achieved, thanks to the following reasons: first, the 1/f noise of the tail current source can appear as a CM signal at T_1 primary and modulate the oscillation voltage. However, the T_1 transformer will effectively filter out this CM AM signal, thus preventing any AM-to-PM conversion at the C_2 switched capacitors and nonlinear C_{gs} of gm-devices. Second, the class-F$_2$ tank has two impedance peaks at the fundamental oscillation frequency and its second harmonic. Hence, the second harmonic of the drain current flows into a resistance of the tank instead of its capacitive part. It effectively reduces the 1/f noise upconversion to the $1/f^3$ phase noise due to Groszkowski phenomenon [24]; we will discuss this phenomenon intensively in Chapter 5. Third, the soft clipping effectively reduces the voltage variation of V_T, as shown in Figure 4.11. Intuitively, it could reduce the DC and even-order coefficients of ISF at this node and thus alleviate the 1/f noise conversion of the tail current transistor.

The PN noise beyond the 10-MHz offset is dominated by thermal noise floor from the divider and buffer set at −162 dBc/Hz. The oscillator has a 19% tuning range from 7.2 to 8.7 GHz. Figure 4.17 shows the phase noise and FoM of the oscillator at 3-MHz offset across the tuning range (after the ÷2 divider). The average FoM is as high as 191 dBc/Hz and varies less than 2 dB. The oscillator also reveals a very low frequency pushing of 42 and 22 MHz/V at the highest and lowest frequencies, respectively.

Figure 4.18 shows the PN performance versus its current consumption. The circuit cannot satisfy Barkhausen oscillation criterion at $I_{DC} < 7$ mA. The oscillator phase noise is improved only by 10 dB/dec between 7 and 12 mA due to the drop in the oscillator current efficiency α_I and loading of the tank's Q-factor by the gm-devices entering the linear region. Note

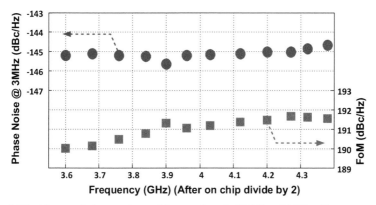

Figure 4.17 Measured phase noise and figure of merit (FoM) at 3 MHz offset versus carrier frequency.

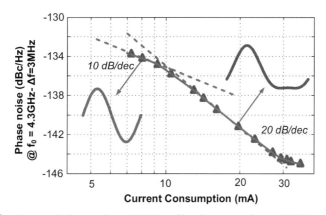

Figure 4.18 Measured phase noise at 3 MHz offset frequency from 4.3 GHz carrier versus the oscillator current consumption.

that even though the tank has an additional impedance at $2\omega_0$, the second harmonic of the drain current is negligible and, consequently, the drain oscillation resembles a sinusoid. However, by further increasing the drain current, the soft clipping phenomenon appears where the tank loading and tail transistor noise effects are reduced significantly due to the class-F_2 operation. Consequently, PN improves by almost 20 dB/dec, which demonstrates a few dB of improvement compared to the traditional class-B operation (compare Figures 4.3 and 4.18). Figure 4.18 also indicates that the circuit can sustain the oscillation even with $4\times$ lower I_{DC} and thus G_{MEF}, which translates into sufficient margin for the oscillator start-up over PVT variations.

Table 4.3   Comparison with relevant ultra-low phase noise oscillators

	This work	[13]	[27]	[1]	[17]	[25]	[26]
Technology	CMOS 65 nm	CMOS 65 nm	CMOS 65 nm	CMOS 65 nm	MOS 350 μm	CMOS 55 nm	BiCMOS 130 nm
Supply voltage (V)	1.3	2.15	1.5	1.25	2.5	1.5	3.3
Frequency (GHz)	4.35[1]	4.07	3.92[1]	3.7[1]	1.2	3.35[1]	1.56
Tuning range (%)	19	19	10.2	25	18	31.4	9.6
PN at 3 MHz (dBc/Hz)	−144.8	−146.7	−147.7	−142.2	−152	−142	−150.4
Norm. PN[2] (dBc/Hz)	−158.3	−159.6	−157.2	−154.3	−154.8	−153.3	−155
Power consumption (mW)	41.6	126.8	48	15	9.25	27	290
FoM (dB)	191.8[3]	188.3	190.1	192.2	195	189	180
FoM_T[4] (dB)	197.4	193.4	190.3	200.2	200.7	199	179.7
Transformers/ inductors count	2	2	2	1	2	1	1
Oscillator structure	Class-F₂	Dual core Class-C	Hard clipping	Class-F₃	Noise filtering	Class-B/ Class-C	Colpitts

[1] After on-chip ÷2 divider.
[2] At 3-MHz offset frequency normalized to 915-MHz carrier.
[3] FoM drops to 191.5 dB by considering the divider power consumption of 2.6 mW.
[4] $FOM_T = |PN| + 20 \log_{10}((f_0/\Delta f)\,(TR/10)) - 10 \log_{10}(P_{DC}/1mW)$.

Table 4.3 summarizes the performance of the class-F₂ oscillator and compares it with the best spectral purity relevant oscillators. Note that this oscillator demonstrates the best PN with the highest power efficiency at a relatively low supply voltage. Only the dual-core class-C oscillator [13] offers better PN performance but at the price of $1.65\times$ larger V_{DD}, $3\times$ higher power consumption, and 3 dB lower FoM or power efficiency.

4.6 Conclusion

In this chapter, we have presented and analyzed a class-F₂ oscillator where an auxiliary impedance peak is introduced around the second harmonic of the oscillating waveform. The second harmonic of the active device current converts into voltage and, together with the fundamental component, creates a soft clipped oscillation waveform. The class-F_2 operation offers enough headroom for the low noise operation of the tail current transistor without compromising the oscillator current and voltage efficiencies. Furthermore, the special ISF of the soft clipping waveform reduces significantly the circuit-to-phase-noise conversion. The additional resonant frequency is realized by exploiting a different transformer behavior in common-mode and differential-mode excitations. In addition, the tank input impedance is also scaled down without sacrificing its Q-factor. Consequently, this structure is able to push the phase noise much lower than practically possible with the traditional LC oscillators while satisfying long-term reliability requirements.

References

[1] M. Babaie and R. B. Staszewski, "A class-F CMOS Oscillator," *IEEE J. Solid-State Circuits*, vol. 48, no. 12, pp. 3120–3133, Dec. 2013.

[2] R. B. Staszewski and P. T. Balsara, *All-Digital Frequency Synthesizer in Deep-Submicron CMOS. Wiley*, 2006. Available: http://books.google.nl/books?id=2VHFD-7LgAwC.

[3] C. Weltin-Wu, G. Zhao, and I. Galton, "A 3.5 GHz digital fractional-N PLL frequency synthesizer based on ring oscillator frequency-to-digital conversion," *IEEE J. Solid-State Circuits*, vol. 50, no. 12, pp. 2988–3002, Dec. 2015.

[4] L. Vercesi, L. Fanori, F. D. Bernardinis, A. Liscidini, and R. Castello, "A dither-less all digital PLL for cellular transmitters," *IEEE J. Solid-State Circuits*, vol. 47, no. 8, pp. 1908–1920, Aug. 2012.

[5] K. Takinami, R. Strandberg, P. C. P. Liang, G. L. G. de Mercey, T. Wong, and M. Hassibi, "A rotary-traveling-wave-oscillator-based alldigital PLL with a 32-phase embedded phase-to-digital converter in 65 nm CMOS," *IEEE International Solid-State Circuits Conference Digest of Technical Papers (ISSCC)*, 2011, pp. 100–101.

[6] C. Hsu, M. Z. Straayer, and M. H. Perrott, "3.6 GHz digital $\Delta\Sigma$ fractional-N frequency synthesizer with a noiseshaping time-to-digital converter and quantization noise cancellation," *IEEE J. Solid-State Circuits*, vol. 43, no. 12, pp. 2776–2786, Dec. 2008.

[7] H. H. Chang, P.-Y. Wang, J.-H. Zhan, and H. Bing-Yu, "A fractional spur-free ADPLL with loop-gain calibration and phase-noise cancellation for GSM/GPRS/EDGE," *IEEE International Solid-State*

Circuits Conference Digest of Technical Papers (ISSCC), 2008, pp. 200–201.

[8] M. Babaie, A. Visweswaran, Z. He, and R. B. Staszewski, "Ultra-low phase noise 7.2–8.7 GHz clip-and-restore oscillator with 191 dBc/Hz FoM," *Proceedings of IEEE Radio Frequency Integrated Circuits (RFIC) Symposium*, 2013, pp. 43–46.

[9] M. Babaie and R. B. Staszewski, "An ultra-low phase noise class-F$_2$ CMOS oscillator with 191 dBc/Hz FOM and long term reliability," *IEEE J. Solid-State Circuits*, vol. 50, no. 3, pp. 679–692, Mar. 2015.

[10] R. B. Staszewski, M. Babaie, and Z. He, "Oscillator," *US Patent 9,337,847*, issued 10 May 2016.

[11] M. Babaie and R. B. Staszewski, "An ultra-low phase noise class-F$_2$ CMOS oscillator with 191 dBc/Hz FOM and long term reliability," *IEEE J. Solid-State Circuits*, vol. 50, no. 3, pp. 679–692, Mar. 2015.

[12] L. Romano, A. Bonfanti, S. Levantino, C. Samori, and A. L. Lacaita, "5-GHz oscillator array with reduced flicker up-conversion in 0.13-μm CMOS," *IEEE J. Solid-State Circuits*, vol. 41, no. 11, pp. 2457–2467, Nov. 2006.

[13] M. Tohidian, S. Mehr, and R. B. Staszewski, "Dual-core high-swing class-C oscillator with ultra-low phase noise," *Proceedings of IEEE Radio Frequency Integrated Circuits (RFIC) Symposium*, 2013, pp. 243–246.

[14] J. R. Long, "Monolithic transformers for silicon RF IC design," *IEEE J. Solid-State Circuits*, vol. 35, no. 9, pp. 1368–1382, Sept. 2000.

[15] A. Mazzanti and P. Andreani, "Class-C harmonic CMOS VCOs, with a general result on phase noise," *IEEE J. Solid-State Circuits*, vol. 43, no. 12, pp. 2716–2729, Dec. 2008.

[16] L. Fanori and P. Andreani, "Highly efficient class-C CMOS VCOs, including a comparison with class-B VCOs," *IEEE J. Solid-State Circuits*, vol. 48, no. 7, pp. 1730–1740, Jul. 2013.

[17] E. Hegazi, H. Sjoland, and A. A. Abidi, "A filtering technique to lower LC oscillator phase noise," *IEEE J. Solid-State Circuits*, vol. 36, no. 12, pp. 1921–1930, Dec. 2001.

[18] P. Andreani, X. Wang, L. Vandi, and A. Fard, "A study of phase noise in Colpitts and LC-tank CMOS oscillators," *IEEE J. Solid-State Circuits*, vol. 40, no. 5, pp. 1107–1118, May 2005.

[19] J. Chen, L. Ye, D. Titz, F. Gianesello, R. Pilard, A. Cathelin, F. Ferrero, C. Luxey, and A. Niknejad, "A digitally modulated mm-Wave cartesian beamforming transmitter with quadrature spatial combining,"

IEEE International Solid-State Circuits Conference Digest of Technical Papers (ISSCC), 2013, pp. 232–233.

[20] A. M. Niknejad and R. G. Meyer, "Analysis, design, and optimization of spiral inductors and transformers for Si RF ICs," *IEEE J. Solid-State Circuits*, vol. 33, no. 10, pp. 1470–1481, Oct. 1998.

[21] A. Hajimiri and T. H. Lee, "A general theory of phase noise in electrical oscillators," *IEEE J. Solid-State Circuits*, vol. 33, no. 2, pp. 179–194, Feb. 1998.

[22] D. Murphy, J. J. Rael, and A. A. Abidi, "Phase noise in LC oscillators: A phasor-based analysis of a general result and of loaded Q," *IEEE Transactions on Circuits and Systems I, Reg. Papers*, vol. 57, no. 6, pp. 1187–1203, Jun. 2010.

[23] C.-H. Jan, M. Agostinelli, H. Deshpande, M. El-Tanani, W. Hafez, U. Jalan, L. Janbay, M. Kang, H. Lakdawala, J. Lin, Y.-L. Lu, S. Mudanai, J. Park, A. Rahman, J. Rizk, W.-K. Shin, K. Soumyanath, H. Tashiro, C. Tsai, P. Vandervoorn, J.-Y. Yeh, and P. Bai, "RF CMOS Technology Scaling in High-k/Metal Gate Era for RF SoC (Systemon-Chip) Applications," *IEEE International Electron Devices Meeting (IEDM)*, 2010, pp. 604–607.

[24] J. Rael and A. Abidi, "Physical processes of phase noise in differential LC oscillators" *Proceedings of IEEE Custom Integrated Circuits Conference (CICC)*, 2000, pages: 569–572.

[25] L. Fanori, A. Liscidini, and P. Andreani, "A 6.7-to-9.2 GHz 55 nm CMOS hybrid class-B/class-C cellular TX VCO," in *IEEE International Solid-State Circuits Conference Digest of Technical Papers (ISSCC)*, 2012, pp. 354–355.

[26] J. Steinkamp, F. Henkel, P. Waldow, O. Pettersson, C. Hedenas, and B. Medin, "A Colpitts oscillator design for a GSM base station synthesizer," *Proceedings of IEEE Radio Frequency Integrated Circuits (RFIC) Symposium*, 2004, pp. 405–408.

[27] A. Visweswaran, R. B. Staszewski, and J. R. Long, "A low phase noise oscillator principled on transformer-coupled hard limiting," *IEEE J. Solid-State Circuits*, vol. 49, no. 2, pp. 300–311, Feb. 2014.

5

A 1/f Noise Upconversion Reduction Technique

In this chapter, we introduce a method to reduce a flicker (1/f) noise upconversion in voltage-biased RF oscillators. Excited by a harmonically rich tank current, a typical oscillation voltage waveform is observed to have asymmetric rise and fall times due to even-order current harmonics flowing into the capacitive part, as it presents the lowest impedance path. The asymmetric oscillation waveform results in an effective impulse sensitivity function (ISF) of a non-zero dc value, which facilitates the 1/f noise upconversion into the oscillator's $1/f^3$ phase noise. We demonstrate that if the ω_0 tank exhibits an auxiliary resonance at $2\omega_0$, thereby forcing this current harmonic to flow into the equivalent resistance of the $2\omega_0$ resonance, then the oscillation waveform would be symmetric and the flicker noise upconversion would be largely suppressed. The auxiliary resonance is realized at no extra silicon area in both inductor- and transformer-based tanks by exploiting different behavior of inductors and transformers in differential- and common-mode excitations. These tanks are ultimately employed in designing modified class-D and class-F oscillators in 40-nm CMOS technology. They exhibit an average flicker noise corner of less than 100 kHz.

5.1 Introduction

Close-in spectra of RF oscillators are degraded by a flicker (1/f) noise upconversion. The resulting low-frequency phase noise (PN) fluctuations can be mitigated as long as they fall within a loop bandwidth of an enclosing phase-locked loop (PLL). However, the PLL loop bandwidths in cellular transceivers are less than a few tenths to a few hundreds of kHz [1,2], which is below the typical $1/f^3$ PN corner of CMOS oscillators [3–5]. Consequently, a

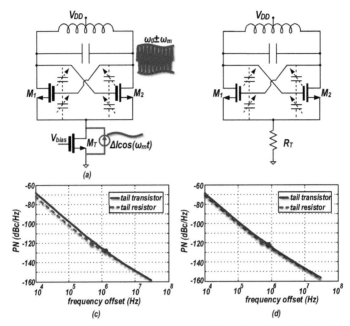

Figure 5.1   Class-B oscillator: (a) with tail transistor M_T; (b) with tail resistor R_T; and their PN when (c) M_T is always in saturation; (d) M_T enters partially into triode.

considerable amount of the oscillator's low frequency noise cannot be filtered by the loop and will adversely affect the transceiver operation.

In a current-biased oscillator, flicker noise of a tail transistor, M_T, modulates the oscillation voltage amplitude and then upconverts to PN via an AM–PM conversion mechanism through nonlinear parasitic capacitances of active devices, varactors, and switchable capacitors [6,7] (see Figure 5.1(a)).[1] An intuitive solution is to configure the oscillator into a *voltage-biased* regime, which involves removing the M_T [8] or replacing it with a tail resistor, R_T, in Figure 5.1(b). Such expected reduction is highly dependent on the tail transistor's operating region. If M_T in Figure 5.1(a) is always in saturation, the amount of 1/f noise is considerable and the tail resistor R_T in Figure 5.1(b) could improve the low-frequency PN performance, as shown in Figure 5.1(c). However, in advanced CMOS process nodes with a reduced supply voltage, M_T partially enters the triode region, thereby degrading the oscillator's effective noise factor but improving the 1/f noise

---

[1]It is shown in [6] that for certain values of varactor bias voltages, this upconversion is almost eliminated.

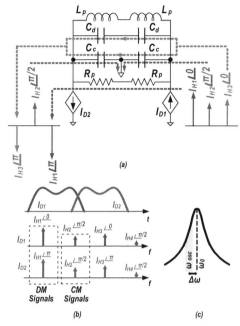

Figure 5.2 (a) Current harmonic paths; (b) drain current in time and frequency domains; (c) frequency drift due to Groszkowski effect.

upconversion; see Figure 5.1(d). In [3], class-C oscillators were designed with a tail transistor and a tail resistor. Measured $1/f^3$ corners are almost the same, thus supporting our discussion.[2] However, regardless of the M_T operating region, removing this source would still not completely eliminate the 1/f noise upconversion.

Another mechanism of the 1/f upconversion is due to Groszkowski effect [10]. In a harmonically rich tank current, the fundamental component, I_{H1}, flows into the equivalent parallel resistance of the tank, R_p. Other components, however, mainly take the capacitive path due to their lower impedance; see Figure 5.2(a). In any balanced RF oscillating circuit, odd harmonics circulate in a differential path, while even harmonics flow in a common-mode path through the resonator capacitance and the switching transistors to ground [32]. Compared to the case with only the fundamental component, the capacitive reactive energy increases by the higher harmonics flowing into them. This phenomenon makes the tank's reactive energy un-balanced.

---

[2]The actual flicker noise reduction mechanism of class-C oscillators was revealed in [9].

The oscillation frequency will shift down from the tank's natural resonance frequency, ω_0, in order to increase the inductive reactive energy and restore the energy equilibrium of the tank. This frequency shift is given by [11]

$$\frac{\Delta\omega}{\omega_0} = -\frac{1}{Q^2}\sum_{n=2}^{\infty}\frac{n^2}{n^2-1}\cdot\left|\frac{I_{Hn}}{I_{H1}}\right|^2,\qquad(5.1)$$

where I_{Hn} is the nth harmonic component of the tank's current. The literature suggests that this shift is static but any fluctuation in I_{Hn}/I_{H1} due to the 1/f noise modulates $\Delta\omega$ and exhibits itself as $1/f^3$ PN [12]; see Figure 5.2(c). Although this mechanism has been known for quite some time, it is still not well understood how the flicker noise modifies the I_{Hn}/I_{H1} ratio. Furthermore, (5.1) suggests that all harmonics *indiscriminately* modulate the Groszkowski's frequency shift by roughly the same amount, without regard to their odd/even-mode nature, which could be easily misinterpreted during the study of the flicker noise upconversion in cross-coupled oscillators.

While recognizing the Groszkowski's frequency shift as the dominant physical mechanism in voltage-biased oscillators, we turn our attention to the impulse sensitivity function (ISF) theory in researching the above questions. Hajimiri and Lee [13] have shown that upconversion of any flicker noise source depends on the dc value of the related effective ISF, which can be significantly reduced if the waveform has certain symmetry properties [13, 14]. Another explanation was offered in [15, 16] suggesting that if the 1/f noise current of a switching MOS transistor is to be modeled by a product of stationary noise and a periodic function $w(t)$, then this noise can upconvert to PN if $w(t)$ is asymmetric.

In this chapter, we elaborate on a method proposed in [22, 23, 34] to effectively trap the second current harmonic into a resistive path of a tank in a *voltage-biased* oscillator topology. Doing so will reduce the core transistors' low frequency noise upconversion by making the oscillation waveform symmetric and reducing the effective ISF dc value. We further investigate the effects of harmonics on the core transistors' flicker noise upconversion by studying their impact on the oscillation waveform and on the effective impulse sensitivity function, $\Gamma_{eff,dc}$.

It should be mentioned that several solutions are proposed in literature to reduce the 1/f noise upconversion due to Groszkowski's frequency shift. The concept of a harmonically rich tank current degrading the close-in oscillator spectrum has been noticed for quite some time; however, the proposed solutions mostly include linearization of the system to reduce the level of current harmonics by limiting the oscillation amplitude by an AGC [17, 18], or linearization of gm-devices [19, 20], at the expense of the oscillator's start-up margin and increased $1/f^2$ PN. In a completely different strategy, a resistor is added in [21] in series with gm-device drains. An optimum value of the resistor minimizes the flicker noise upconversion; however, the 1/f noise improvement is at the expense of the 20-dB/dec degradation in oscillators with low V_{DD} and high current consumption.

The rest of this chapter is organized as follows. Section 5.2 shows how harmonic components of the drain current contribute to the flicker noise upconversion and shows how an auxiliary CM resonance at $2\omega_0$ mitigates this upconversion. Section 5.3 demonstrates how the auxiliary resonance is realized and proves the effectiveness of the proposed method by implementing two classes of voltage-biased oscillators. Section 5.4 reveals the details of circuit implementations and measurement results.

5.2 Method to Reduce 1/f Noise Upconversion

5.2.1 Auxiliary Resonant Frequencies

Let us start by focusing on reducing the Groszkowski frequency shift. As shown in Figure 5.2(a), the oscillation frequency ω_{osc} fluctuates around the tank's natural resonant frequency ω_0 due to the flow of higher harmonics of the current $I_{D1,2}$ into the capacitive part of the tank. A voltage-biased class-B tank current in time and frequency domains is shown in Figure 5.2(b). Odd harmonics of the tank current are differential-mode (DM) signals; hence, they can flow into both differential- and single-ended capacitors. Even harmonics of the tank current, on the other hand, are common-mode (CM) signals and can only flow into single-ended (SE) capacitors. If the tank possesses further resonances coinciding with these higher harmonics (see Figure 5.3(a)), these components can find their respective resistive path to flow into, as shown in Figure 5.3(b). Consequently, the capacitive reactive energy would not be disturbed and the oscillation frequency shift $\Delta\omega$ would be minimized (see Figure 5.3(c)). The input impedance Z_{in} of such a tank is shown in Figure 5.3(d). The tank has the fundamental natural resonant frequency at ω_0 and auxiliary CM and DM resonant frequencies at even- and odd-order harmonics, respectively. Minimizing the frequency shift $\Delta\omega$ will weaken the underlying mechanism of the 1/f noise upconversion; however, realizing auxiliary resonances at higher harmonics has typically been area inefficient and

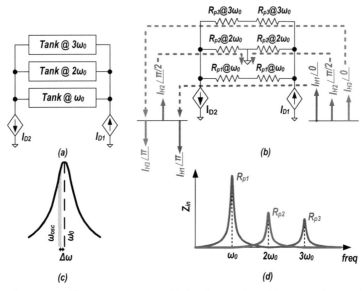

Figure 5.3 (a) Auxiliary resonances at higher harmonics; (b) current harmonic paths; (c) frequency drift; (d) input impedance of the tank.

can also degrade the PN performance. Consequently, the auxiliary resonance frequencies have to be chosen wisely.

Groszkowski frequency shift formula (5.1) indicates that all the contributing current harmonics I_{Hn} are weighted by almost the same coefficients. This means that, in practice, stronger current harmonics I_{Hn} contribute more to the frequency shift. Consequently, we can narrow down the required auxiliary resonances to these harmonics. On the other hand, ultimately, the low frequency noise upconversion depends on the oscillation waveform and the dc value of effective ISF. The various current harmonics contribute unevenly to the flicker noise upconversion since they result in different oscillation waveforms and effective ISF values. Investigating these differences reveals how many and at which frequencies the auxiliary resonances should be realized.

5.2.2 Harmonic Effects on the Effective ISF

A (hypothetical) sinusoidal resonance tank current $I_{H1}(t) = |I_{H1}| \sin(\omega_0 t)$ would result in a sinusoidal resonance oscillation voltage: $V_{H1}(t) = R_{p1} \cdot |I_{H1}| \sin(\omega_0 t) = A_1 \sin(\omega_0 t)$. Its ISF is also a zero-mean sinusoid but

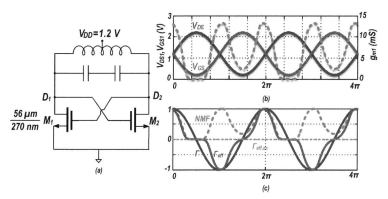

Figure 5.4 Oscillator example: (a) schematic; (b) V_{DS}, V_{GS}, and g_m of M_1 transistor when oscillation voltage contains only fundamental component; (c) its ISF, NMF, and effective ISF.

in quadrature with $V_{H1}(t)$ [24]. The flicker noise of core transistors (e.g., $M_{1,2}$ in Figure 5.4(a)) in a cross-coupled oscillator is modeled by a current source between the source and drain terminals and exhibits a power spectral density as

$$\overline{i_n^2(t)} = \frac{K}{WLC_{ox}} \cdot \frac{1}{f} \cdot g_m^2(\omega_0 t), \tag{5.2}$$

where K is a process-dependent constant, W and L are core transistors' width and length, respectively, and C_{ox} is an oxide capacitance per area. Due to the dependency of current noise on g_m, the flicker noise source is a cyclostationary process and can be expressed as

$$i_n(t) = i_{n0}(\omega_0 t) \cdot \alpha(\omega_0 t), \tag{5.3}$$

in which $i_{n,0}(\omega_0 t)$ shows the stochastic stationarity. $\alpha(\omega_0 t)$ is the noise modulating function (NMF), which is normalized, deterministic, and periodic with maximum of 1. It describes the noise amplitude modulation; consequently, it should be derived from the cyclostationary noise characteristics [13]. In this case, an *effective* impulse sensitivity function is defined as $\Gamma_{eff}(\omega_0 t) = \alpha(\omega_0 t) \cdot \Gamma(\omega_0 t)$. $M_{1,2}$ flicker noise cannot upconvert to PN if effective ISF has a zero dc value.

Let us investigate the $M_{1,2}$ flicker noise upconversion when the oscillation voltage ideally contains only the fundamental component. In Figure 5.4(a), $V_{D1} = V_{DD} - A_1 sin(\omega_0 t)$, $V_{G1} = V_{D2} = V_{DD} + A_1 sin(\omega_0 t)$. Assuming $V_{DD} = 1.2\,\text{V}$ and $A_1 = 1\,\text{V}$, the g_m of the M_1 transistor under such V_{DS} and V_{GS} is found by simulations and is shown as dotted line in Figure 5.4(b).

Under this condition, $\alpha(\omega_0 t) = \frac{g_m(\omega_0 t)}{g_{m,max}}$. ISF, NMF, and the effective ISF of the M_1 flicker noise source are shown in Figure 5.4(c). The dc value of such an effective ISF is zero, resulting in no flicker noise upconversion. This is a well known conclusion and is referred to as a state where $M_{1,2}$ transistors' flicker noise cannot be upconverted to PN [16].

In reality, the tank current of voltage-biased oscillators is rich in harmonics. Due to physical circuit constraints, the even-order current harmonics lead by $\pi/2$, while the odd-order current harmonics are in-phase with the fundamental current I_{H1}. The $\pi/2$ phase difference in even- and odd-order current harmonics considerably changes the oscillation waveform characteristics. For simplicity, we focus only on dominant harmonics, $I_{H2} = |I_{H2}| \sin(2\omega_0 t + \pi/2)$ and $I_{H3} = |I_{H3}| \sin(3\omega_0 t)$, as representatives of even- and odd-order current harmonics, respectively; however, the following discussion can be easily generalized for all harmonics. We also assume for now that the tank only contains SE capacitors.

The differential current I_{H2} flows into the SE capacitors and creates a second-order voltage harmonic:

$$V_{H2}(t) = \frac{1}{C \cdot 2\omega_0} \cdot |I_{H2}| \sin\left(2\omega_0 t + \pi/2 - \pi/2\right) = \alpha_2 A_1 \sin\left(2\omega_0 t\right),$$

(5.4)

where the $-\pi/2$ phase shift is due to the capacitive load. The oscillation voltage will then be

$$V_{T2}(t) = V_{H1}(t) + V_{H2}(t) = A_1 \left[\sin\left(\omega_0 t\right) + \alpha_2 \sin\left(2\omega_0 t\right)\right].$$ (5.5)

$V_{H1}(t)$, $V_{H2}(t)$, and $V_{T2}(t)$ are plotted in Figure 5.5(a) for $\alpha_2 = 0.1$ and $A_1 = 1\,\text{V}$. $V_{H1}(t)$ has two zero-crossings within its period: at t_1 and t_2, their rise and fall times are symmetric with derivatives: $V'_{H1}(t_1) = -V'_{H1}(t_2)$. V_{H2}'s zero-crossings are also at t_1 and t_2; however, $V'_{H2}(t_1) = V'_{H2}(t_2)$. Consequently, the opposite slope polarities of V_{H1} and V_{H2} at t_1 slow down the fall time of V_{T2}, while the same slope polarities at t_2 sharpen its rise time. Consequently, as can be gathered from Figure 5.5(a), V_{T2} features *asymmetric* rise and fall slopes.

The resulting ISF of the gm transistor is calculated based on (36) in Ref. [13] and is shown in Figure 5.5(b), with its mean dependent on α_2. Larger α_2 leads to more asymmetry between $V_{T2}(t)$ rise and fall slopes; hence, $\Gamma_{eff,dc}$ will increase. Furthermore, repeating the same simulations to obtain g_{m1} with drain and gate voltages that contain second harmonic components results in asymmetric g_{m1} and, consequently, NMF. The slower

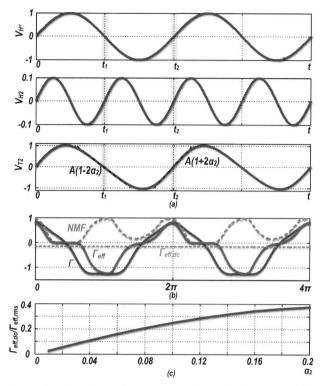

Figure 5.5 Conventional tank waveforms: (a) fundamental, V_{H1}, second harmonic, V_{H2}, voltage components, and oscillation waveform, V_{T2}; (b) its ISF, NMF, and effective ISF; (c) $\Gamma_{eff,dc}/\Gamma_{eff,rms}$ for different α_2 values.

rise/fall times increase the duration when M_1 is turned on, thus widening g_{m1}. A sharper rise/fall time decreases the amount of time when M_1 is turned on, resulting in a narrower g_{m1}. The NMF and effective ISF of such waveforms are shown in Figure 5.5(b). The effective ISF has a dc value which results in $M_{1,2}$'s flicker to PN upconversion. Dependency of the dc value of the effective ISF on α_2 is shown in Figure 5.5(c).

This argument is valid for all even-order current harmonics, and we can conclude that the fluctuations in the even harmonics of the tank's current convert to the $1/f^3$ PN noise through the modulation of the oscillating waveform.

Let us now investigate a case of the tank current containing only odd-harmonic components, with $I_{H3} = |I_{H3}|\sin(3\omega_0 t)$ as a representative. I_{H3}

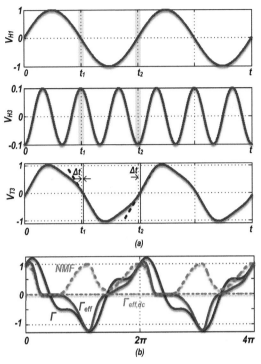

Figure 5.6 Conventional tank waveforms: (a) fundamental, V_{H1}, third harmonic, V_{H3}, voltage component, and oscillation waveform, V_{T3}; (b) its ISF, NMF, and effective ISF.

flows mainly into the tank capacitors and creates a third harmonic voltage as

$$V_{H3}(t) = \frac{1}{C \cdot 3\omega_0} \cdot |I_{H3}| \sin(3\omega_0 t - \pi/2) = \alpha_3 A_1 \sin(3\omega_0 t - \pi/2) \quad (5.6)$$

where, again, the $-\pi/2$ phase shift is due to the capacitive load. The oscillation voltage will then be

$$V_{T3}(t) = V_{H1}(t) + V_{H3}(t) = A_1 [\sin(\omega_0 t) + \alpha_3 \sin(3\omega_0 t - \pi/2)]. \quad (5.7)$$

$V_{H1}(t)$, $V_{H3}(t)$, and $V_{T3}(t)$ are plotted in Figure 5.6(d) for $\alpha_3 = 0.1$ and $A_1 = 1$ V. It is obvious that the oscillation waveform falling and rising slopes are symmetric, and $\Gamma_{dc} = 0$, as easily gathered from Figure 5.6(e). The simulations show that g_{m1} is slightly asymmetric due to amplitude distortion of the oscillation voltage. However, this asymmetry is canceled out when multiplied by ISF (see Figure 5.6(e)), resulting in an effective ISF with almost zero dc value and thus preventing low-frequency noise upconversion. These

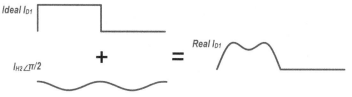

Figure 5.7 Ideal and real current waveforms.

arguments can be generalized for all odd-order harmonics. Consequently, the low-frequency noise of gm transistors does not upconvert to PN if the tank current only contains odd harmonics.

So far, we have assumed $\pi/2$ phase for I_{H2} and π phase for I_{H3}; however, the exact phase shift between the fundamental and harmonic components depends on the transconductor nonlinearity. The ideal drain current in a class-B oscillator is a square wave. However, the core transistors enter triode region, resulting in the real current shape to exhibit a dimple. For this current waveform to appear, a current harmonic with twice the fundamental frequency and $\pi/2$ phase shift has to be added to the original waveform (which only contains odd harmonics), as shown in Figure 5.7. Hence, the phase delay of the second harmonic is not arbitrary but is constrained by the physical circuit. It is worth mentioning that in a class-C oscillator, where the transistors do not enter the triode region, the phase difference between the first and second harmonic is not $\pi/2$. The class-C oscillator shows less $1/f^3$ corner compared to the other topologies which is in agreement with our claim about the importance of the fundamental and second current phase shift on the low frequency noise upconversion. Let us investigate what happens to the voltage waveform and ISF if the fundamental and third-harmonic components are not in-phase, and have a phase shift of ϕ_1. Following the same approach as in the manuscript and assuming $A_1 = 1$, $V_{H3} = \alpha_3 \sin(3w_0 t - \pi/2 + \phi_1)$. If for $\theta_1 = w_0 t_1$, $V_T = V_{H1} + V_{H3} = 0$, then for $\theta_2 = \pi + \theta_1$, $V_T = \sin(\theta_2) + \alpha_3 \sin(3\theta_2 + \phi_1) = -\sin(\theta_1) - \alpha_3 \sin(3\theta_1 + \phi_1) = 0$. Then the V_T slopes at θ_1 and θ_2 would be

$$V_T'(\theta_1) = \cos(\theta_1) + 3\alpha_3 \cos(3\theta_1 + \phi_1)$$
$$V_T'(\theta_2) = \cos(\pi + \theta_1) + 3\alpha_3 \cos(3\pi + 3\theta_1 + \phi_1) = -V_T'(\theta_1). \quad (5.8)$$

Consequently, regardless of the exact phase difference between the current's fundamental and third-harmonic components, the oscillation voltage in the presence of third (and, in general, odd) harmonics would have symmetric

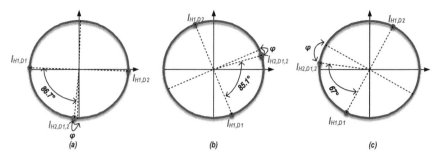

Figure 5.8 Tank's current fundamental and second-harmonic phases in (a) class-D; (b) class-F_3; and (c) class-C topologies.

rise and fall times, and consequently they will not be responsible for 1/f to phase noise upconversion.

For the second-harmonic component, the story is entirely different. The most asymmetric rise and fall times of the oscillation voltage waveform happen when the fundamental and second-harmonic current components have exactly the $\pi/2$ phase difference. However, if the phase difference is not exactly $\pi/2$ (and it is, for example, $\pi/2 - \phi_1$), rise and fall times are still asymmetric and the ISF still contains a non-zero dc value. The proposed method would be still effective not at exactly $\omega_{CM} = 2\omega_0$, but at a CM resonance that has the $\phi_1 - \pi/2$ phase at $2\omega_0$ and cancels out the extra phase difference, ϕ_1, therefore, resulting in a completely symmetric rise and fall times. It is worth mentioning that if ϕ_1 were originally close to $\pi/2$, the oscillation waveform should have theoretically very symmetric rise and fall times. However, due to the real-world circuit constraints, that situation could not be reproduced in simulations.

Independent from the above explanations, we did some circuit-level simulations to show exact phase shift of the second harmonic compared to the fundamental components in class-D, class-F_3, and also class-C topologies. The results are shown in Figure 5.8. For class-D and class-F_3 topologies, the $\pi/2$ phase difference appears to be a very good estimation. However, as predicted for the class-C topology, $\pi/2$ phase difference is not a precise estimation.

To further support that 1/f noise upconverts more to PN if α_2 is increased, we tried to run some simulations on the voltage-biased class-B oscillator of Figure 5.9(a). Controlling the second-harmonic current is not very straightforward. It can be modified by changing the core transistors' width, W, or by changing tank's quality factor, Q. In both of these methods, the oscillation

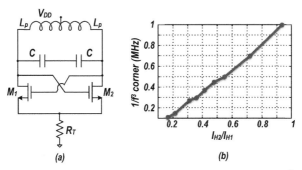

Figure 5.9 (a) Voltage-biased class-B oscillator schematics; (b) $1/f^3$ corner versus I_{H2}/I_{Hn}.

waveform amplitude would get affected. If we fix the oscillation amplitude when W or Q are swept, the second harmonic power modification range becomes very limited. On the other hand, by changing the transistors' width, the flicker noise of the device also changes and adds another parameter. Consequently, in the following simulations, we swept the tank quality factor, and all the other parameters, such as transistor sizes, supply voltage, etc., are kept the same. With higher Q, the oscillation voltage increases, the device spends more time in triode region and becomes more non-linear, consequently generating more current harmonics. M_1 and M_2 in the class-B oscillator are thick-oxide 56μ/270n devices, $V_{DD} = 1.2V$, $R_T = 9$ Ohm. The capacitors are ideal and not tunable. The simulation results are shown in Figure 5.9(b). As we have discussed the flicker noise upconversion depends on the α_2 value, which is proportional to the I_{H2}/I_{Hn} ratio. Therefore, we reported the $1/f^3$ corner versus I_{H2}/I_{Hn}, and it is obvious from Figure 5.9(b) that the corner increases for larger I_{H2}/I_{Hn} ratios in this class-B oscillator.

5.2.3 Resonant Frequency at $2\omega_0$

Thus far, we have shown that the even components of the tank's current are chiefly accountable for the asymmetric oscillation waveform and the 1/f noise upconversion to PN. Let us investigate what happens to the oscillation waveform and effective ISF if the tank has an auxiliary CM resonance at $2\omega_0$. Such resonance provides a resistive (i.e., via R_{p2}) path for I_{H2} to flow into it, and hence the voltage second-harmonic component is

$$V_{H2,aux}(t) = R_{p2}|I_{H2}|\sin\left(2\omega_0 t + \pi/2\right) = A_1\alpha_{2,aux}\sin\left(2\omega_0 t + \pi/2\right).$$
$$(5.9)$$

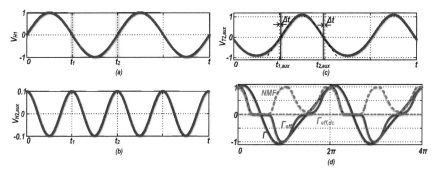

Figure 5.10   Proposed tank waveforms: (a) fundamental voltage component, V_{H1}; (b) voltage second harmonic in the presence of auxiliary resonance, $V_{H2,aux}$; (c) oscillation waveform, $V_{T2,aux}$; (d) its ISF, NMF, and effective ISF.

The composite oscillation voltage will become

$$V_{T2,aux}(t) = V_{H1}(t) + V_{H2,aux}(t)$$
$$= A_1 \left[\sin(\omega_0 t) + \alpha_{2,aux} \sin(2\omega_0 t + \pi/2) \right]. \quad (5.10)$$

$V_{H1}(t)$, $V_{H2,aux}(t)$, and $V_{T2,aux}(t)$ are plotted in Figure 5.10(a,b,c) for $\alpha_{2,aux} = 0.1$ and $A_1 = 1$. The rise and fall times of the oscillation voltage are now symmetric (see Figure 5.10(c)) and so the ISF is zero mean, as shown in Figure 5.10(d). g_{m1}, and thus NMF, are also completely symmetrical; consequently, the effective ISF has a zero dc value, preventing low-frequency noise from being upconverted. The oscillation waveform is still dependent on $\alpha_{2,aux}$, but the rise and fall times are always symmetric, thus keeping $\Gamma_{eff,dc}$ zero.

The second and third current harmonics are the most dominant in all classes of oscillators, so α_2 and α_3 are significantly larger than other α_n for $n = 4, 5, \ldots$. Meanwhile, Γ_{dc} is a growing function of α_n for $n = 2k$, where $k = 1, 2, \ldots$. We can, therefore, conclude that I_{H2} is the main contributor to the 1/f noise upconversion. Consequently, attention to only one auxiliary resonant frequency at $2\omega_0$ appears sufficient [22, 26, 27].

5.2.4 ω_{CM} Deviation from $2\omega_0$

The balance in the rise and fall zero-crossing slopes in Figure 5.10(c) is rooted in the $\pi/2$ phase shift between $V_{H1}(t)$ and $V_{H2}(t)$. This is a combination of the $\pi/2$ phase difference between $I_{H1}(t)$ and $I_{H2}(t)$ and zero phase of the resistive tank impedance at $2\omega_0$. When ω_{CM} deviates from $2\omega_0$

$$
\begin{aligned}
V_{T2,aux}(t) &= V_{H1}(t) + V_{H2,aux}(t) \\
&= R_{p1}|I_{H1}|\sin(\omega_0 t) + |Z_{CM}| \cdot |I_{H2}|\sin(2\omega_0 t + \pi/2 + \phi_{CM}) \\
&= A_1\left[\sin(\omega_0 t) + \alpha_{2,aux}\sin(2\omega_0 t + \pi/2 + \phi_{CM})\right] \quad (5.11)
\end{aligned}
$$

where $|Z_{CM}|$ and ϕ_{CM} are the CM input impedance magnitude and phase, respectively, derived as,

$$
\phi_{CM} = \arctan\left(\frac{1 - \zeta^2}{\frac{\zeta}{Q_{CM}}}\right) \quad (5.12)
$$

$$
|Z_{CM}| = R_{p2} \cdot \frac{\frac{\zeta}{Q_{CM}}}{\sqrt{(1 - \zeta^2)^2 + \left(\frac{\zeta}{Q_{CM}}\right)^2}} \quad (5.13)
$$

where $\zeta = \frac{2\omega_0}{\omega_{CM}}$. The ω_{CM} versus $2\omega_0$ misalignment has two effects. The first directly translates ϕ_{CM} into the waveform asymmetry. Figure 5.11(a) shows $V_{T2,aux}(t)$ for different ϕ_{CM}; $\alpha_{2,aux}$ was chosen as 0.3 to better illustrate the asymmetry. When grossly mistuned from $2\omega_0$, ϕ_{CM} could approach $\pm\pi/2$, thus making the auxiliary resonance completely ineffective. A larger Q-factor of the common-mode resonance, Q_{CM}, results in ϕ_{CM} closer to $\pm\pi/2$ for the same $2\omega_0/\omega_{CM}$ ratios, as illustrated in Figure 5.11(b).

The second effect is due to $\alpha_{2,aux}$, which determines the amount of second harmonic in the voltage waveform. When ϕ_{CM} is not zero, $\Gamma_{eff,dc}$ becomes dependent on $\alpha_{2,aux}$: the larger $\alpha_{2,aux}$, the more asymmetric waveform and more 1/f noise upconversion. The $\alpha_{2,aux}$ value can be found from the following equation:

$$
\alpha_{2,aux} = \left|\frac{I_{H2}}{I_{H1}}\right| \cdot \frac{|Z_{CM}|}{R_{p1}}. \quad (5.14)
$$

I_{H2}/I_{H1} is dependent on the oscillator's topology. Furthermore, the larger Q_{CM}, the larger R_{p2} and hence the larger $\alpha_{2,aux}$. Figure 5.11(c) shows the expected $\Gamma_{eff,dc}/\Gamma_{eff,rms}$ versus ϕ_{CM} for different $\alpha_{2,aux}$. Both of these effects point out that Q_{CM} should be low to reduce the sensitivity of this method to the ω_{CM} deviation from $2\omega_0$. Parsitic inductances and capacitances between supply and groud rails can contribute to this deviation. This parasitics become especially important at mmW frequencies [25].

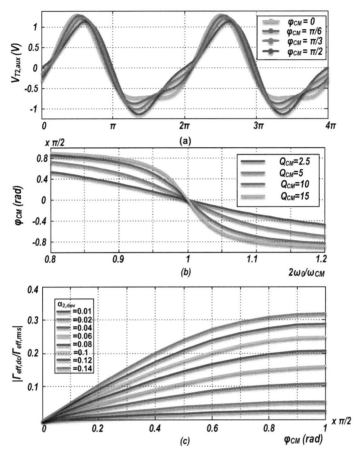

Figure 5.11   (a) $V_{T2,aux}$ for different ϕ_{CM}; (b) ϕ_{CM} for different Q_{CM} when ω_{CM} deviates from $2\omega_0$; (c) $\Gamma_{eff,dc}/\Gamma_{eff,rms}$ for different $\alpha_{2,aux}$ and ϕ_{CM}.

5.3 Circuit Implementation

We have shown that if the tank demonstrates an auxiliary CM resonance at the second harmonic of its fundamental ω_0 resonance, the oscillation waveform would be symmetric and, hence, the flicker noise upconversion would be suppressed. Since the differential capacitors are not seen by the CM signals (i.e., I_{H2}), a straightforward solution for realizing a CM peak is to design a tank as demonstrated in Figure 5.12(a)) with a set of differential C_d and single-ended (SE) C_c capacitors [26, 27]. r_p is the equivalent series resistance of the inductor and it is assumed that all capacitors are nearly ideal. This tank

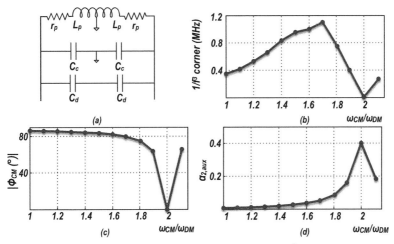

Figure 5.12 (a) A tank with DM and CM resonances; (b) $1/f^3$ corner of the oscillator employing this tank; (c) ϕ_{CM}; and (d) $\alpha_{2,aux}$ of the tank versus ω_{CM}/ω_{DM}.

shows a fundamental DM resonant frequency, $\omega_{DM} = \dfrac{1}{\sqrt{L_p(C_c+C_d)}}$ and a CM resonant frequency $\omega_{CM} = \dfrac{1}{\sqrt{L_pC_c}}$. From (5.12)–(5.14):

$$\phi_{CM} = \arctan\left(\frac{1-\frac{4C_c}{C_c+C_d}}{\frac{1}{Q_{DM}}\cdot\frac{2C_c}{C_c+C_d}}\right) \tag{5.15}$$

$$\alpha_{2,aux} = \frac{R_{p2}}{R_{p1}}\cdot\frac{\frac{2}{Q_{DM}}\cdot\left(\frac{C_c}{C_c+C_d}\right)}{\sqrt{\left(1-\frac{4C_c}{C_c+C_d}\right)^2+\left(\frac{2}{Q_{DM}}\cdot\frac{C_c}{C_c+C_d}\right)^2}}\cdot\frac{I_{H2}}{I_{H1}}, \tag{5.16}$$

where Q_{DM}, R_{p2}, and R_{p1} are, respectively, the quality factor at DM resonance, and impedance peaks at CM and DM resonances. In an extreme condition of $C_d = 0$, the tank contains only the SE capacitors and reduces to a conventional tank discussed in Section 5.2.2. Targeting $\omega_{CM} = 2\omega_{DM}$ results in $C_d = 3C_c$ and we can prove that $Q_{CM} = 2Q_{DM}$. As discussed supra, the fairly large Q_{CM} exacerbates the effects of CM resonance misalignment.

To investigate the effectiveness of the proposed method on the tank mistuning sensitivity, we performed an analysis of a 5-GHz voltage-biased class-B oscillator of Figure 5.1(b) with $Q_{DM} = 10$. The oscillator is designed in a 40-nm CMOS technology, and $M_{1,2}$ are thick-oxide (56/0.27)-μm devices. The power consumption is 10.8 mW at $V_{DD} = 1.2$ V. As expected, the $1/f^3$ corner of this oscillator is at its minimum of $\sim 10\,\text{kHz}$ at $C_d/C_c = 3$ (see Figure 5.12(b)). When ω_{CM} deviates from $2\omega_{DM}$, i.e., C_d/C_c ratio deviates from the ideal value of 3, while keeping $C_c + C_d$ constant, the $1/f^3$ corner starts to increase from the $10\,\text{kHz}$ minimum and reaches its peak at $\omega_{CM} = 1.7\omega_{DM}$ when the CM resonance phase, ϕ_{CM}, gets close to $\pi/2$ (about $80°$ as shown in Figure 5.12(c)). After this point, the ϕ_{CM} barely changes, but $\alpha_{2,aux}$ decreases (Figure 5.12(d)) and, consequently, the $1/f^3$ corner reduces again. The maximum $1/f^3$ corner of $1.1\,\text{MHz}$ is actually much worse than the $400\,\text{kHz}$ corner of extreme case when $C_d = 0$ (see Figure 5.12(b)). This means that if the tank is not designed properly, the performance would be even worse than that without applying this technique. Consequently, to ensure no performance degradation in face of the misalignment, $\alpha_{2,aux}$ at $\phi_{CM} \approx 80°$ should be less than that of the tank without the applied technique. α_2 when $C_d = 0$ can be found from (5.16):

$$\alpha_2 \approx \frac{2}{3Q_{DM}} \cdot \frac{I_{H2}}{I_{H1}} \tag{5.17}$$

$\phi_{CM} = 80°$, (5.15) and (5.16) result in

$$\alpha_{2,aux} = \frac{R_{p2}}{R_{p1}} \cdot \frac{\tan\left(\frac{\pi}{18}\right)}{\sqrt{1 + \tan^2\left(\frac{\pi}{18}\right)}} \cdot \frac{I_{H2}}{I_{H1}}. \tag{5.18}$$

Hence,

$$\frac{R_{p2}}{R_{p1}} < \frac{3.84}{Q_{DM}} \tag{5.19}$$

to satisfy this condition, which results in non-practical Q_{DM} values.

In the following two subsections, we show how to substantially reduce the sensitivity to such misalignment by employing, at no extra area penalty, an inductor exhibiting distinct and beneficial characteristics in DM and CM excitations. The different behavior of a 1:2 turn transformer in DM and CM excitations is also exploited to design a transformer-based F_2 tank. With these new tanks, we construct class-D and a class-F oscillators to demonstrate the effectiveness of the proposed method of reducing the flicker noise upconversion. Before we do that, let us compare the $1/f^3$ corner and current harmonics of a current-biased class-B (see Figure 5.13(a)), a current-biased class-B with noise filtering technique [32] applied to it (see Figure 5.13(b)), and a voltage-biased class-B with the proposed technique applied to it (see Figure 5.13(c)). In the current-biased configuration proposed in [32], a capacitor in parallel with the current source shorts noise frequencies around $2\omega_0$

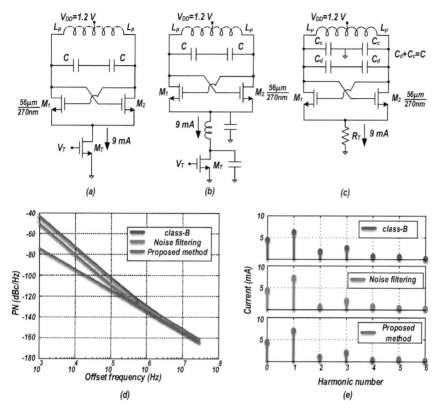

Figure 5.13 (d) PN and (e) current harmonic components of a class-B oscillator (a), and similar counterparts with noise filtering technique [32] (b) and the proposed method applied (c).

to ground. An inductor is interposed between the common source of the cross-coupled transistors and the current source. That inductor resonates at $2\omega_0$ with the equivalent capacitor at the common source of the oscillator transistors (see Figure 5.13(b)). The purpose of that resonator is creating a high impedance path at $2\omega_0$ to stop the tank loading when one of the core transistors enters the triode region. That method is also partially effective in reducing low-frequency noise upconversion of core transistors by linearizing the core transistors and reducing current harmonics, especially the second-harmonic component content. That resonator has to be tunable over the tuning range and it increases the die area. Figures 5.13(d–e) show the PN performance and current harmonic components of the three oscillators shown in Figure 5.13(a–c). In the simulations, all the capacitors are ideal capacitors and non-tunable. It is obvious how much the second-harmonic current is reduced in the noise filtering method, and consequently the $1/f^3$ corner is lower than that in a conventional class-B oscillator.

5.3.1 Inductor-Based F_2 Tank

Figures 5.14(a,b) show a 2-turn "F_2" inductor when it is excited by DM and CM signals. In the DM excitation, currents in both turns have the same direction, resulting in an additive magnetic flux. However, in the CM excitation, currents have opposite direction and cancel each other's flux [28]. With the proper spacing between the F_2 inductor windings, effective inductance in CM can be made 4x smaller than that in DM. The L_{DM}/L_{CM} inductance ratio is controlled through lithography that *precisely* sets the physical inductor dimensions and, consequently, makes it *insensitive* to process variations.

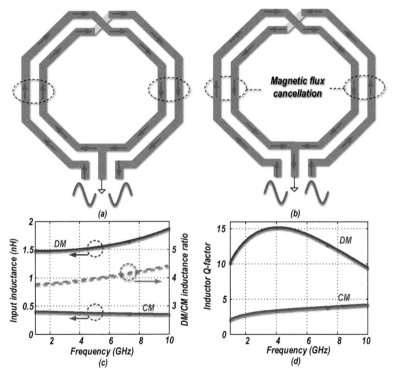

Figure 5.14   A 2-turn "F_2" inductor in (a) DM excitation; (b) CM excitation; (c) F_2 DM and CM inductances and their ratio; (d) Q_{DM} and Q_{CM}.

Figure 5.14(c) shows the DM and CM inductances and their ratio over frequency. L_{DM}/L_{CM} is close to 4 within a 30%–40% tuning range.

Differential capacitors cannot be seen by the CM signals; hence, to be able to set the CM resonance, the F_2 tank capacitors should be SE, as shown in Figure 5.15(a). The F_2 tank demonstrates two resonant frequencies: ω_{DM} and ω_{CM}. Both of these are tuned simultaneously by adjusting C_c. The precise inductor geometry maintains $L_{DM}/L_{CM} \approx 4$ and hence $\omega_{CM}/\omega_{DM} \approx 2$ over the full tuning range.

The input impedance of the tank is shown in Figure 5.15(b). Presuming the capacitance losses are negligible, the DM and CM resonance quality factors are

$$Q_{DM} = \frac{L_p\omega_{DM}}{r_p} = Q_0 \tag{5.20}$$

$$Q_{CM} = \frac{L_p\omega_{CM}}{4r_p} = \frac{Q_0}{2}. \tag{5.21}$$

The Q-factor of the CM resonance is half that of DM, which relaxes the F_2 tank sensitivity to mismatch between ω_{CM} and $2\omega_{DM}$. For this inductor-based F_2 tank, $R_{p2}/R_{p1} = 0.25$ and the condition in (5.19) is *satisfied* for $Q_0 < 15$. Furthermore, in the CM excitation, the currents in adjacent windings have opposite direction, which results in an increased AC resistance [29] and so the Q-factor of the CM inductance is even smaller than in (5.21). The Q-factor of L_{CM} inductance of Figure 5.14(b) is about 3–4.

Apart from the easy tuning with only one capacitor bank, the mostly SE parasitic capacitors do not play any role in defining the $\omega_{CM}/2\omega_{DM}$ ratio.

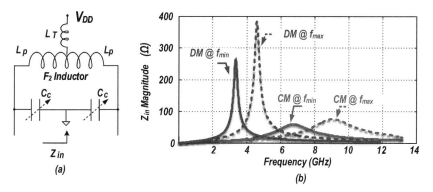

Figure 5.15 (a) Inductor-based F_2 tank and (b) its input impedance.

Furthermore, the low Q_{CM} and, consequently, the lower sensitivity to the $\omega_{CM}/2\omega_{DM}$ ratio that the inductor-based F_2 tank offers make it all more attractive than the tank shown in Figure 5.12(a).

5.3.2 Class-*D*/F_2 Oscillator

Among the various classes of inductor-based oscillators (e.g., class-B, complementary class-B, class-D [4]), we have decided to validate the proposed method on a class-D oscillator depicted in Figure 5.16(a). This recently introduced oscillator shows promising PN performance in the $1/f^2$ region due to its special ISF. The tail current transistor is removed there and wide and almost ideal switches $M_{1,2}$ clip the oscillation voltage to GND for half a period (see Figure 5.16(b)) resulting in an almost zero ISF there

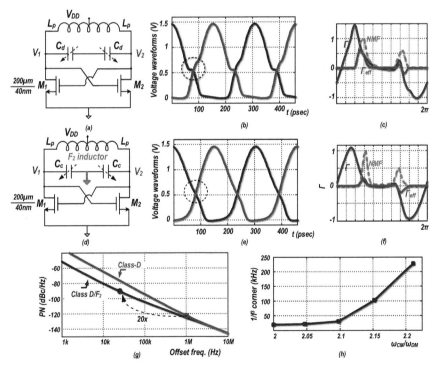

Figure 5.16 Class-D oscillator: (a) schematic; its (b) waveforms; and (c) gm-transistor ISF, NMF, and effective ISF. Class-D/F_2 oscillator: (d) schematic; its (e) waveforms; and (f) gm-transistor ISF, NMF, and effective ISF; (g) their PN performance; and (h) $1/f^3$ corner sensitivity to ω_{CM}/ω_{DM}.

(Figure 5.16(c)). However, the hard clipping of the drain nodes to GND generates a huge amount of higher-order harmonic currents. Due to the large I_{H2}, in agreement with our analysis, the oscillating waveform has asymmetric fall and rise times (clearly visible in Figure 5.16(b)) and it exhibits a strong 1/f noise upconversion and frequency pushing. A version of class-D with a tail filter technique [32] was also designed in [4] in an attempt to reduce the low frequency noise upconversion. This method is partially effective, lowering the $1/f^3$ PN corner from 2 to 0.6–1 MHz. Due to the above reasons, this voltage-biased oscillator seems a perfect fit for the proposed method.

Figure 5.16(d) shows the proposed class-D/F_2 oscillator, which adopts the F_2 tank. The gm-devices, M_1 and M_2, still inject a large I_{H2} current into the tank, but this current is now flowing into the equivalent resistance of the tank at $2\omega_0$. Clearly, the rise/fall times are more symmetric in the class-D/F_2 oscillator, as demonstrated in Figure 5.16(e). The gm-transistors' ISF, NMF, and effective ISF are shown in Figure 5.16(f). As predicted, effective Γ_{dc} is now reduced and the simulated PN performance shows that the $1/f^3$ corner is lowered from 1 MHz to ~30 kHz (Figure 5.16(g)).

The parasitic inductance L_T has to be considered in designing the F_2 inductor. C_c controls both CM and DM resonant frequencies simultaneously; hence, any deviation of ω_{CM} from $2\omega_0$ is due to L_{CM}/L_{DM} not being exactly 4 over the TR. To examine the robustness of the tank via simulations, a C_d differential capacitor is deliberately added to the tank. $C_c + C_d$ is kept constant in order to maintain the oscillation frequency. This capacitor shifts down ω_{DM} while keeping ω_{CM} intact. Figure 5.16(h) shows how the $1/f^3$ corner worsens when C_d/C_c ratio increases. The class-D/F_2 oscillator is quite robust to process variations. First of all, due to the low CM resonance quality factor of an F_2 inductor, this topology is less sensitive to deviations of ω_{CM} from $2\omega_0$. Furthermore, in this topology, only single-ended capacitor banks are employed; hence, ω_{CM}/ω_{DM} ratio is solely defined by L_{CM}/L_{DM}. The cross-coupled transistor's parasitic capacitors are mostly single-ended, except for C_{gd} which is less than 5% of the tank's total capacitance. Consequently, the modification of core transistors' parasitic capacitance over process variations will only change oscillation frequency and barely modify the ω_{CM}/ω_{DM} ratio. Figure 5.17(a) shows simulation results for the $1/f^3$ corner of the class-D/F_2 oscillator in different process corners. The PN at 10-kHz offset frequency for both class-D/F_2 oscillators for 200 points Monte Carlo simulations on inter/intra die process variations is shown in Figure 5.17(a).

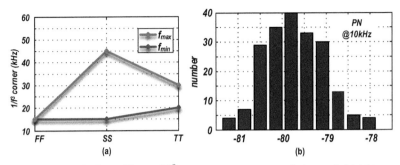

(a)   (b)

Figure 5.17   Class-D/F_2 oscillator: $1/f^3$ corner over process variation and (b) histogram of PN at 10-kHz offset frequency.

5.3.3 Transformer-Based F_2 Tank

Figures 5.18(a,b) show a 1:2 turns transformer excited by DM and CM input signals at its primary. With a DM excitation, the induced currents at the secondary circulate in the same direction leading to a strong coupling factor, k_m. On the other hand, in CM excitation, the induced currents cancel each other, resulting in a weak k_m [30]. The latter means that the secondary winding cannot be seen by the CM signals. "F_2" transformer-based tank is shown in Figure 5.19(a).

At the DM excitation, no current flows into the metal track inductance, L_T, that connects the center tap to the supply's AC-ground

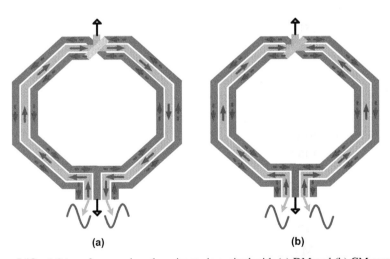

(a)   (b)

Figure 5.18   1:2 transformer when the primary is excited with (a) DM and (b) CM currents.

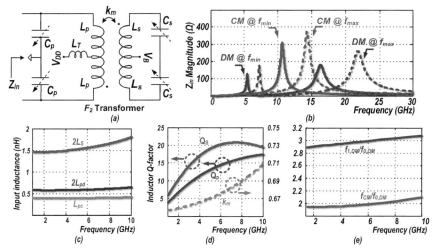

Figure 5.19 (a) Transformer-based F_2 tank; (b) its input impedance; (c) DM and CM primary and secondary inductance; (d) primary and secondary inductance quality factor and coupling factor; (e) DM and CM resonant frequencies over TR.

(see Figure 5.19(a)). However, at the CM excitation, the current flowing into L_T is twice the current circulating in the inductors. Consequently, the tank inductance L_p in Fig. 5.19(a) is re-labeled as $L_{pd} = L_p$ in DM and $L_{pc} = L_{pd} + 2L_T$ in CM excitations. This tank employs SE primary and differential secondary capacitors and demonstrates two DM and one CM resonant frequencies. $\omega_{CM} = 1/\sqrt{L_{pc}C_p}$ and if $k_{m,DM} > 0.5$, $\omega_{0,DM} = 1/\sqrt{L_{pd}C_p + L_sC_s}$ [5]. F_2 tank requires $\omega_{CM} = 2\omega_{0,DM}$; hence,

$$L_sC_s = C_p(4L_{p,c} - L_{p,d}). \tag{5.22}$$

Unlike in the inductor-based tank, here, the $\omega_{CM}/\omega_{0,DM}$ ratio is dependent on the secondary-to-primary capacitor ratio. Furthermore, the input impedance Z_{in}, shown in Figure 5.19(b), reveals that Q_{CM} is not low, thus making it sensitive to $\omega_{CM}/\omega_{0,DM}$. It means the C_s/C_p ratio has to be carefully designed to maintain $\omega_{CM}/\omega_{0,DM} \approx 2$ over the tuning range. In practice, the Q-factor of capacitor banks is finite and decreases at higher frequencies, so Q_{CM} will reduce, thus making the tank a bit less sensitive.

5.3.4 Class-$F_{2,3}$ Oscillator

As proven in [5], a DM auxiliary resonance at the third harmonic of the fundamental frequency is beneficial in improving the 20-dB/dec PN performance

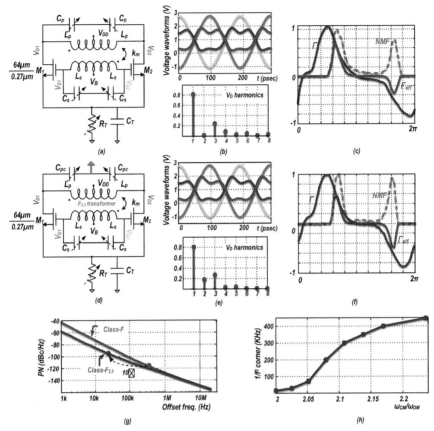

Figure 5.20   Class-F_3 oscillator: (a) schematic; (b) its waveforms; and (c) gm-transistor ISF, NMF, and effective ISF. Class-$F_{2,3}$ oscillator: (d) schematic; (e) its waveforms; and (f) gm-transistor ISF, NMF, and effective ISF; (g) their PN performance; and (h) $1/f^3$ corner sensitivity to ω_{CM}/ω_{DM}.

by creating a pseudo-square-wave oscillation waveform (see Figure 5.20(b)). We can merge our transformer-based F_2 tank with the class-F_3 operation in [5] to design a class-$F_{2,3}$ oscillator, as shown in Figure 5.20(d–e). To ensure $\omega_{CM} = 2\omega_{0,DM}$ and $\omega_{1,DM} = 3\omega_{0,DM}$, we force $L_sC_s = 3.8L_{pd}C_p$ and $k_m = 0.67$. The relatively low k_m increases the impedance at $\omega_{1,DM} \equiv 3\omega_{0,DM}$ [31]. However, the class-F_3 oscillator meets the oscillation criteria only at $\omega_{0,DM}$. Figure 5.20(e) demonstrates that the pseudo-square waveform of class-F_3 oscillation is preserved in the class-$F_{2,3}$ oscillator. The waveform does not appear to differ much; however, the oscillation voltage spectrum indeed confirms the class-$F_{2,3}$ operation. I_{H2} is not very large in this class of oscillators, consequently, the fall/rise-time asymmetry is not as distinct as in the class-D oscillator. However, the $1/f^3$ corner improvement from 400 kHz in class-F_3 to <30 kHz in class-$F_{2,3}$, as demonstrated in Figure 5.20(g), proves the effectiveness of the method. The ISF, NMF, and effective ISFs for these oscillators are shown in Figure 5.20(c,f).

Class-$F_{2,3}$ oscillator performance is sensitive to the deviation of ω_{CM} from $2\omega_0 \equiv 2\omega_{DM}$. C_p changes both CM and DM resonant frequencies while C_s only changes the DM one. To examine the robustness of the F_2 operation, differential capacitors are added in the tank's primary. Here again $C_{p,c} + C_{p,d}$ is constant to maintain the oscillation frequency. Figure 5.20(h) shows the $1/f^3$ corner versus. ω_{CM}/ω_{DM} ratio and underscores the need to control the capacitance ratio, as per (5.22). Otherwise, a small deviation increases the $1/f^3$ corner rapidly, and with larger deviations, the method becomes ineffective. The class-$F_{2,3}$ topology is somewhat less robust to process variations. This topology is generally more sensitive to deviations of ω_{CM} from $2\omega_0$ due to higher CM resonance quality factor. Moreover, in this topology, $\omega_{CM}/\omega_{DM} = \sqrt{((L_s C_s + L_p C_p)/(L_p C_p))}$ and is dependent on secondary and primary windings' inductance and capacitance ratio. The modification of core transistors parasitic capacitance modifies both primary and secondary capacitances and, hence, both DM and CM resonant frequencies are affected in the same direction (they are both going to increase with less parasitic capacitance and vice versa). Therefore, the ω_{CM}/ω_{DM} ratio is not affected drastically with the parasitic capacitance variations; however, due to larger Q_{CM}, the small effects are also important. Figure 5.21(a) shows the simulation results for $1/f^3$ corner of the class-$F_{2,3}$ oscillator in different

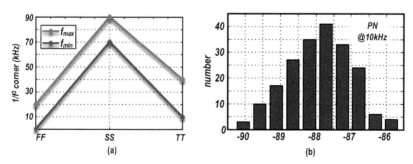

Figure 5.21 Class-$F_{2,3}$ oscillator: $1/f^3$ corner over process variation and (b) histogram of PN at 10-kHz offset frequency.

process corners. The $1/f^3$ corner is more affected compared to the class-D/F_2 oscillator; however, it is still very competitive. The PN at 10-kHz offset frequency for class-$F_{2,3}$ oscillator for 200 points Monte Carlo simulations on inter/intra die process variations is shown in Figure 5.21(b).

It should be worthwhile to mention that recently a mm-wave class-F_{23} oscillator considering the second-harmonic current return path was discussed in [35].

5.4 Experimental Results

The class-D/F_2 and class-$F_{2,3}$ oscillators, whose schematics were shown in Figure 5.16(d) and Figure 5.20(d), respectively, are designed in 40-nm CMOS to demonstrate the suppression of the 1/f noise upconversion. For fair comparison, we attempted to design the oscillators with the same specifications, such as center frequency, tuning range, and supply voltage, as their original reference designs in [4] and [5].

5.4.1 Class-D/F_2 Oscillator

The class-D/F_2 oscillator is realized in a 40-nm 1P8M CMOS process *without* ultra-thick metal layers. The two-turn inductor is constructed by stacking the 1.45 μm Alucap layer on top of the 0.85 μm top (M8 layer) copper metal. The DM inductance is 1.5 nH with simulated Q of 12 at 3 GHz. Combination of MOS/MOM capacitors between the supply and ground is placed on-chip to minimize the effective L_T inductance, and the remaining uncompensated inductance is modeled very carefully. The capacitor bank is realized with 6-bit switchable MOM capacitors with LSB of 30 fF. The oscillator is tunable between 3.3 and 4.5 GHz (31% TR) via this capacitor bank. $M_{1,2}$ transistors are (200/0.04)-μm low-V_t devices to ensure start-up and class-D operation over PVT. The chip micrograph is shown in Figure 5.22(a) with core area of 0.1 mm^2.

Figure 5.23(a) shows the measured PN at f_{max} and f_{min} with $V_{DD} = 0.5$ V. Current consumption is 6 and 4 mA, respectively. The $1/f^3$ corner is 100 kHz at f_{max} and reduces to 60 kHz for f_{min}. The $1/f^3$ corner over TR is shown in Figure 5.23(c). The supply frequency pushing is 60 and 40 MHz/V at f_{max} and f_{min}, respectively (see Figure 5.23(b)). Table 5.1 compares its performance with the original class-D oscillators (as well as two other relevant oscillators [21, 26] aimed at reducing the 1/f noise upconversion).

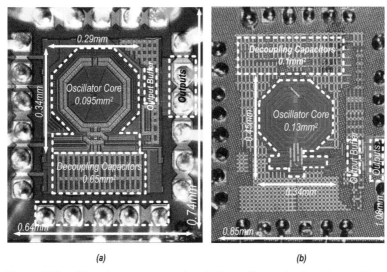

Figure 5.22 Chip micrographs: (a) class-D/F_2 oscillator; (b) class-$F_{2,3}$ oscillator.

Compared to the original design, the FoM at 10-MHz offset is degraded in the class-D/F_2 oscillator by 3 dB, mainly due to the lack of ultra-thick metal layers, which lowers the inductor's Q. However, even with this degradation, FoM at 100-kHz offset is improved at least 3 dB. $1/f^3$ corner is improved at least 10 times versus both class-D and noise-filtering class-D oscillators.

5.4.2 Class-$F_{2,3}$ Oscillator

The class-$F_{2,3}$ oscillator is realized in 40-nm 1P7 CMOS process with ultra-thick metal layer. The 1:2 transformer is constructed with the 3.4 µm top ultra-thick (M7 layer) copper metal. The primary and secondary winding inductances are 0.58 and 1.5 nH, respectively, and $k_m = 0.67$. The simulated Q-factors of the primary and secondary windings are 15 and 20 at 6 GHz. Like the class-D/F_2, the L_T inductance has to be compensated with enough decoupling capacitance. The unfiltered part has to be modeled precisely due to the relatively large R_{p2}. The single-ended primary and differential secondary capacitor banks are realized with two 6-bit switchable MOM capacitors with LSB of 30 fF and 50 fF, respectively. Due to the sensitivity of this oscillator to the frequency mismatch, an 8-bit unit-weighted capacitor bank with LSB of 4 fF is also placed at the primary to tune the DM and CM resonance frequencies. The oscillator is tunable between 5.4

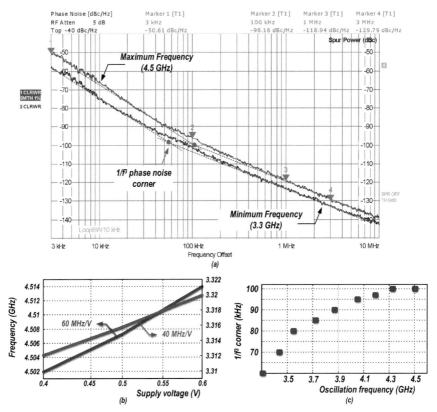

Figure 5.23　Class-D/F$_2$ oscillator: measured (a) PN at $f_{\max}$ and $f_{\min}$; (b) frequency pushing due to supply voltage variation; and (c) $1/f^3$ corner over tuning range.

Table 5.1　Performance summary and comparison with relevant oscillators

		Class-D/F$_2$	Class-D [4]	Noise filtering Class-D [4]	Class-F$_{2,3}$	Class-F$_3$[5]	[21]	[26]	[33]
Technology		**40 nm**	65 nm	65 nm	**40 nm**	65 nm	0.35 μm BiCMOS	65 nm	28 nm
Thick metal		**No**	Yes	Yes	**Yes**	Yes	NA	NA	NA
V$_{DD}$ (V)		**0.5**	0.4	0.4	**1**	1.25	2.7	1.2	0.9
Tuning range (%)		**31**	45	45	**25**	25	14	18	27.2
Core area (mm²)		**0.1**	0.12	0.15	**0.13**	0.12	NA	0.08	0.19
Freq. (GHz)		$f_{\min}$　$f_{\max}$	$f_{\min}$　$f_{\max}$	$f_{\min}$　$f_{\max}$	$f_{\min}$　$f_{\max}$	$f_{\max}$	f_{mid}	$f_{\max}$	f_{mid}
		3.3　**4.5**	3　4.8	3　4.8	**5.4**　**7**	7.4	1.5	3.6	3.3
P$_{DC}$ (mW)		**4.1**　**2.5**	6.8　4	6.8　3.6	**12**　**10**	15	16.2	0.72	6.8
PN (dBc /Hz)	100kHz	**-101.2**　**-96.2**	-101　-91	-102　-92.5	**105.3**　**102.1**	-98.5	-110	-94.4	-106
	1MHz	**-123.4**　**-119**	-127　-119	-128　-121	**126.7**　**124.5**	-125	132	-114.4	-130
	10MHz	**-143.4**　**-139**	-149.5　-143.5	-150　-144.5	**146.7**　**144.5**	-147	NA	-134.5	-150
FoM† (dB)	100kHz	**185.4**　**185.3**	182.2　178.6	183.2　180.6	**189.1**　**188.9**	184.1	181.5	186.8	188.1
	1MHz	**187.6**　**188**	188.2　186.6	189.2　189.1	**190.5**　**191.4**	190.6	183.5	186.9	192.2
	10MHz	**187.6**　**188**	190.7　191.1	191.2　192.6	**190.5**　**191.4**	192.6	NA	187	192.2
$1/f^3$ corner (kHz)		**60**　**100**	800　2100	650　1500	**60**　**130**	700	11	10	200
Freq. pushing (MHz/V)		**40 @0.5 V**　**60 @0.5 V**	140 @0.5 V　480 @0.5 V	90 @0.5 V　390 @0.5 V	**12 @1V**　**23 @1V**	50 @1.25 V	NA	15 @1.2 V	NA

†FOM= $|PN|$ +20 $log_{10}(\omega_0/\Delta\omega)$ -10 $log_{10}(P_{DC}/1mW)$.

and 7 GHz and the primary and secondary capacitors are changed simultaneously to preserve the class-$F_{2,3}$ operation. The $M_{1,2}$ transistors are $(64/0.27)\,\mu m$ thick-oxide devices to tolerate large voltage swings. The tail resistor R_T bank is realized with a fixed 40-Ω resistor in parallel with 7-bit binary-weighted switchable resistors with LSB size of 5 Ω. This bank can tune the oscillation current from 5–20 mA. The chip micrograph is shown in Figure 5.22(b); the core die area is 0.12 mm^2.

Figure 5.24(a) shows the measured PN of the class-$F_{2,3}$ oscillator at f_{max} and f_{min} with $V_{DD} = 1$ V. Current consumption is 10 and 12 mA, respectively. The 1/f^3 corner is 130 kHz at f_{max} and reduces to ∼60 kHz at f_{min}. The 1/f^3 corner over TR is plotted in Figure 5.24(c). The supply frequency pushing is 23 and 12 MHz/V at f_{max} and f_{min}, respectively (see Figure 5.24(b)). Table 5.1 compares its performance with the original class-F_3

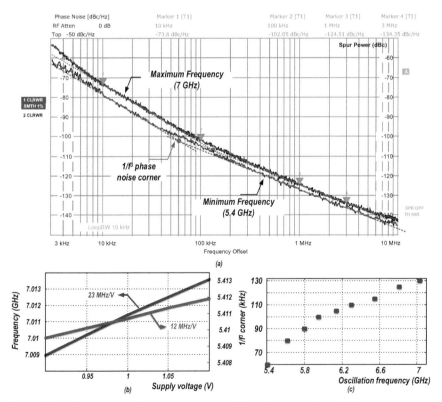

Figure 5.24 Class-$F_{2,3}$ oscillator: measured (a) PN at f_{max} and f_{min}; (b) frequency pushing due to supply voltage variation; and (c) 1/f^3 corner over tuning range.

oscillator. Compared to the original design, FoM is degraded about 1–2 dB at the 10-MHz offset, which is due to the tail resistor loading the tank more than the tail transistor originally, thus degrading PN slightly. Despite this degradation, FoM at 100 kHz is enhanced by at least 4 dB, and the $1/f^3$ corner is improved 5 times.

It might be interesting to point out that class-$F_{2,3}$ oscillator was adapted for an operation at cryogenic temperatures of 4K [36] where it exhibited a record-low 1/f corner.

5.5 Conclusion

In this chapter, we discussed and analyzed a technique to reduce a 1/f noise upconversion in a harmonically rich tank current. We showed that when even-order harmonics of the tank current flow into the capacitive part of the tank, they distort the oscillation waveform by making its rise and fall times asymmetric and hence causing the 1/f noise upconversion. Odd-order harmonics also distort the oscillation waveform; however, the waveform in that case is still symmetric and will not result in the 1/f noise upconversion. We showed how to design a ω_0-tank that shows an auxiliary common-mode (CM) resonant peak at $2\omega_0$, which is the main contributor to the 1/f noise upconversion, and showed how oscillation waveform becomes symmetric by the auxiliary resonance. We described how to realize the F_2-tank without the die area penalty, by taking advantage of different properties of inductors and transformers in differential-mode (DM) and CM excitations. Class-D/F_2 and class-$F_{2,3}$ oscillators employing, respectively, inductor- and transformer-based F_2 tanks are designed in 40-nm CMOS to show the effectiveness of our proposed method. The $1/f^3$ corner improves $10\times$ in class-D/F_2 and $5\times$ in class-$F_{2,3}$ versus their original counterparts.

References

[1] H. Darabi, H. Jensen, and A. Zolfaghari, "Analysis and design of small-signal polar transmitters for cellular applications," *IEEE J. Solid-State Circuits*, vol. 46, no. 6, pp. 1237–1249, June 2011.

[2] D. Tasca et al., "A 2.9-4.0 GHz fractional-N digital PLL with bang-bang phase detector and 560-fs rms integrated jitter at 4.5-mW power," *IEEE J. Solid-State Circuits*, vol. 44, no. 12, pp. 2745–2758, Dec. 2011.

[3] L. Fanori and P. Andreani, "Highly efficient class-C CMOS VCOs, including a comparison with class-B VCOs," *IEEE J. Solid-State Circuits*, vol. 48, no.7, pp. 1730–1740, Jul. 2013.

[4] L. Fanori and P. Andreani, "Class-D CMOS oscillators," *IEEE J. Solid-State Circuits*, vol. 48, no. 12, pp. 3105–3119, Dec. 2013.

[5] M. Babaie and R. B. Staszewski, "A class-F CMOS oscillator," *IEEE J. Solid-State Circuits*, vol. 48, no. 12, pp. 3120–3133, Dec. 2013.

[6] E. Hegazi and A. A. Abidi, "Varactor characteristics, oscillator tuning curves, and AM-FM conversion," *IEEE J. Solid-State Circuits*, vol. 38, no. 6, pp. 1033–1039, Jun. 2003.

[7] B. Soltanian and P. Kinget, "AM-FM conversion by the active devices in MOS LC-VCOs and its effect on the optimal amplitude," *IEEE Radio Frequency Integrated Circuits Symp.*, Jun. 2006.

[8] S. Levantino et al., "Frequency dependence on bias current in 5-GHz CMOS VCO's: Impact of tuning range and flicker noise up-conversion," *IEEE J. Solid-State Circuits*, vol. 37, no. 8, pp. 1003–1011, Aug. 2002.

[9] Y. Hu, T. Siriburanon, and R. B. Staszewski, "Intuitive understanding of flicker noise reduction via narrowing of conduction angle in voltage-biased oscillators", *IEEE Trans. on Circuits and Systems II (TCAS-II)*, pp. 1–5, 2019.

[10] J. Groszkowski, "The impedance of frequency variation and harmonic content, and the problem of constant-frequency oscillator," *Proc. IRE*, vol. 21, pp. 958–981, 1933.

[11] A. Bevilacqua and P. Andreani, "On the bias noise to phase noise conversion in harmonic oscillators using Groszkowski theory," *IEEE Int. Symp. Circuits Syst.*, 2011, pp. 217–220.

[12] J. Rael and A. Abidi, "Physical processes of phase noise in differential LC oscillators," *in Proc. IEEE Custom Integr. Circuits Conf.*, Sept. 2000, pp. 569–572.

[13] A. Hajimiri and T. H. Lee, "A general theory of phase noise in electrical oscillators," *IEEE J. Solid-State Circuits*, vol. 33, no. 2, pp. 179–194, Feb. 1998.

[14] J. E. Post, I. R. Linscott, and M. H. Oslick, "Waveform symmetry properties and phase noise in oscillators," *Electron. Lett.*, vol. 34, no. 16, pp. 1547–1548, Aug. 1998.

[15] D. Murphy, J. J. Rael, and A. A. Abidi, "Phase noise in LC oscillators: A phasor-based analysis of a general result and of loaded Q," *IEEE Trans. Circuits Syst. I, Reg. Papers*, vol. 57, no. 6, pp. 1187–1203, June 2010.

[16] A. Bevilacqua and P. Andreani, "An analysis of 1/f noise to phase noise conversion in CMOS harmonic oscillators," *IEEE Trans. Circuits Syst. I, Reg. Papers*, vol. 59, no. 5, pp. 938–945, May 2012.

[17] E. A. Vittoz, M. G. R. Degrauwe, and S. Bitz, "High-performance crystal oscillator circuits: theory and application," *IEEE J. Solid-State Circuits*, vol. 23, no. 3, pp. 774–783, Jun. 1988.

[18] M. A. Margarit, J. L. Tham, R. G. Meyer, and M. J. Deen, "A low-noise, low-power VCO with automatic amplitude control for wireless applications," *IEEE J. Solid-State Circuits*, vol. 34, no. 6, pp. 761–771, Jun. 1999.

[19] A. R. Jeng and C. G. Sodini, "The impact of device type and sizing on phase noise mechanisms" *IEEE J. Solid-State Circuits*, vol. 40, no. 2, pp. 360–369, Feb. 2005.

[20] S.-J. Yun, C. Y. Cha, H. C. Choi, and S. G. Lee, "RF CMOS LC-oscillator with source damping resistors," *IEEE Microw. Wireless Compon. Lett.*, vol. 16, no. 9, pp. 511–513, Sep. 2006.

[21] F. Pepe, A. Bonfanti, S. Levantino, C. Samori, and A. L. Lacaita, "Suppression of flicker noise up-conversion in a 65-nm CMOS VCO in the 3-to-3.6 GHz band," *IEEE J. Solid-State Circuits*, vol. 48, no. 10, pp. 2375–2389, Oct. 2013.

[22] M. Shahmohammadi, M. Babaie, and R. B. Staszewski, "A 1/f noise up-conversion reduction technique applied to class-D and class-F oscillators," *in IEEE Int. Solid-State Circuits Conf. Dig Tech. Papers (ISSCC)*, Feb. 2015, pp. 444–445.

[23] M. Shahmohammadi, M. Babaie, and R. B. Statszewsi, "A 1/f noise upconversion reduction technique for voltage-biased RF CMOS oscillators," *IEEE J. Solid-State Circuits*, vol. 51, no. 11, pp. 2610–2624, Nov. 2016.

[24] P. Andreani, X. Wang, L. Vandi, and A. Fard, "A study of phase noise in Colpitts and LC-tank CMOS oscillators," *IEEE J. Solid-State Circuits*, vol. 40, no. 5, pp. 1107–1118, May 2005.

[25] Z. Zong, P. Chen, and R. B. Staszewski, "A Low-noise fractional-N digital frequency synthesizer with implicit frequency tripling for mm-Wave applications," *IEEE J. Solid-State Circuits*, vol. 54, no. 3, pp. 755–767, Mar. 2019.

[26] D. Murphy, H. Darabi, and H. Wu, "A VCO with implicit common mode resonance," *in IEEE Int. Solid-State Circuits Conf. Dig Tech. Papers (ISSCC)*, Feb. 2015, pp. 442–443.

[27] D. Murphy, H. Darabi, and H. Wu, "Implicit common-mode resonance in LC oscillators," *IEEE J. Solid-State Circuits*, vol. 52, no. 3, pp. 812–821, Mar. 2017.

[28] J. Chen et al., "A Digitally Modulated mm-Wave Cartesian Beamforming Transmitter with Quadrature Spatial Combining," *in IEEE Int. Solid-State Circuits Conf. Dig Tech. Papers (ISSCC)*, Feb. 2013, pp. 232–233.

[29] D. Chowdhury, L. Ye, E. Alon, and A. M. Niknejad, "An efficient mixed-signal 2.4-GHz polar power amplifier in 65-nm CMOS technology," *IEEE J. Solid-State Circuits*, vol. 46, no. 8, pp. 1796–1809, Aug. 2011.

[30] M. Babaie and R. B. Staszewski, "An ultra-low phase noise class-F2 CMOS oscillator with 191 dBc/Hz FOM and long term reliability," *IEEE J. Solid-State Circuits*, vol. 50, no. 3, pp. 679–692, Mar. 2015.

[31] A. Mazzanti and A. Bevilacqua, "On the phase noise performance of transformer-based CMOS differential-pair harmonic oscillators," *IEEE Trans. Circuits Syst. I, Reg. Papers*, vol. 62, no. 9, pp. 2334–2341, Sep. 2015.

[32] E. Hegazi, H. Sjoland, and A. A. Abidi, "A filtering technique to lower LC oscillator phase noise," *IEEE J. Solid-State Circuits*, vol. 36, no. 12, pp. 1921–1930, Dec. 2001.

[33] A. Mostajeran, M. Sharif Bakhtiar, and E. Afshari, "A 2.4 GHz VCO with FOM of 190 dBc/Hz at 10kHz-to-2MHz offset frequencies in 0.13 μm CMOS using an ISF manipulation technique," *in IEEE Int. Solid-State Circuits Conf. Dig Tech. Papers (ISSCC)*, Feb. 2015, pp. 452–453.

[34] M. Shahmohammadi, M. Babaie, and R. B. Staszewski, "Resonator circuit," *US Patent Application 2017/0366137*, published 21 Dec. 2017.

[35] Y. Hu, T. Siriburanon, and R. B. Staszewski, "A low-flicker-noise 30-GHz class-F_{23} oscillator in 28-nm CMOS using implicit resonance and explicit common-mode return path," *IEEE Journal of Solid-State Circuits (JSSC)*, vol. 53, no. 7, pp. 1977–1987, July 2018.

[36] B. Patra, R. M. Incandela, J. P. G. van Dijk, H. A. R. Homulle, L. Song, M. Shahmohammadi, R. B. Staszewski, A. Vladimirescu, M. Babaie, F. Sebastiano, and E. Charbon, "Cryo-CMOS circuits and systems for quantum computing applications," *IEEE Journal of Solid-State Circuits (JSSC)*, vol. 53, no. 1, pp. 309–321, Jan. 2018.

6

A Switching Current-Source Oscillator

Ultra-low-power (ULP) transceivers underpin short-range communications for wireless Internet-of-Things (IoT) applications. However, their system lifetime is extremely limited by the transceiver power consumption and available battery technology. On the other hand, energy harvesting technologies typically deliver supply voltages that are much lower than the standard supply of CMOS circuits; e.g., on-chip solar cells can supply only 200–800 mV. Although boost converters can bring the level up to the required ∼1 V, their poor efficiency (≤80%) wastes the harvested energy. Consequently, RF oscillators, as one of the transceiver's most power hungry circuitry, must be very power efficient and preferably operate directly at the energy harvester output. In this chapter, we analyze in depth design of an oscillator topology to address the aforementioned constraints without sacrificing manufacturability and phase purity.

6.1 Introduction

Ultra-low-power (ULP) radios underpin short-range communications for wireless Internet-of-Things (IoT). Since RF transmitters (TX) have consumed a significant portion, if not the majority, of the radio's power, the IoT system lifetime tends to be severely limited by the TX power consumption and available battery technology.

Figure 6.1 shows the system lifetime for various battery choices as a function of current consumption. State-of-the-art Bluetooth Low Energy (BLE) radios [1, 2] consume ∼7 mW and thus can *continuously* operate no more than 40 hours on an SR44 battery, which has a comparable dimension to the radio module. This directly causes inconvenient battery replacements at least every few months, which limits their attractiveness from the market

123

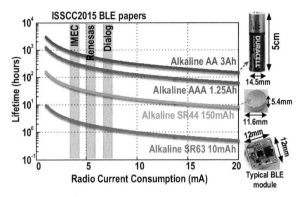

Figure 6.1   BLE system lifetime versus radio current consumption for various battery types.

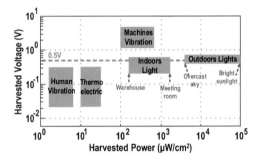

Figure 6.2   Delivered voltage and power density for various harvester types.

perspective. The lifetime can be easily increased by employing larger batteries, but that comes at a price of increased weight and dimensions and it is clearly against the miniaturization vision of IoT. This has motivated an intensive research leading to miniaturized transceivers with a high power efficiency [1–12].

Energy harvesting from a surrounding environment can enable and further spur the IoT applications by significantly extending their lifetime. The delivered voltage versus power density of different harvesting methodologies is depicted in Figure 6.2 [10, 13]. Solar cells offer the highest harvested power per area in both indoor and outdoor conditions. However, they provide lower voltages (0.25–0.75 V) than the expected deep-nanoscale CMOS supply of ∼1 V. Hence, boost converters are typically used to bring the supply level up to the required ∼1 V. As can be gathered from Table 6.1, the relatively poor efficiency (≤80%) of boost converters wastes the harvested energy, thus worsening the system-level efficiency, in addition to increasing

Table 6.1 Performance summary of state-of-the-art boost converters

	[14]	[15]	[16]
	ISSCC'12	ISSCC'14	ISSCC'15
Technology	N/A	65 nm CMOS	0.18 μm CMOS
Input voltage range	0.1–2.9 V	0.15–0.5 V	0.45–3 V
Output voltage range	3 V	0.5–0.6 V	3.3 V
Efficiency @ $V_{in} = 0.5$ V	$\leq 80\%$	$\leq 72.5\%$	$\leq 78.5\%$

the hardware complexity coupled with issues of switching ripples. Consequently, it would be highly desirable for the ULP transceivers to operate directly from the harvested voltage.

The rest of this chapter is dedicated to introduce and analyze a switching current-source oscillator [17, 18] which is optimized for 28-nm CMOS, can operate directly at the low voltage of harvesters, and reduces power and supply voltage without compromising the robustness of the oscillator start-up or loading its tank quality factor.

6.2 Oscillator Power Consumption Trade-offs

The oscillator phase noise (PN) requirements can be calculated by considering the toughest BLE blocking profile:

$$\mathcal{L}(\Delta\omega) = P_{signal} - P_{blocker} - SNR_{min} - 10log_{10}(BW). \qquad (6.1)$$

The received packet error rate (PER) must be better than 30.8% while the wanted signal is just 3 dB above the reference sensitivity level of –70 dBm and in face of an in-band blocker of –40 dBm located at 3-MHz offset from the desired channel. Furthermore, the required signal-to-noise ratio (SNR) should be better than 15 dB to support such PER for a GFSK signal with a modulation index $m = 0.5$ [20]. By replacing the aforementioned values in (6.1), PN shall be better than –105 dBc/Hz at $\Delta\omega = 2\pi \cdot 3$ MHz. Hence, the PN requirements are quite trivial for IoT applications and can be easily met by LC oscillators as long as Barkhausen start-up criterion is satisfied over process, voltage, and temperature (PVT) variations.[1] Consequently, reducing oscillator power consumption, P_{DC}, is the ultimate goal in IoT applications.

---

[1]Ring oscillators can also satisfy such a relaxed phase noise requirement. However, they consume much higher power than LC oscillators at $f_0 \geq 1$ GHz [19].

The PN of any class of an RF oscillator (i.e., class-B) at an offset frequency $\Delta\omega$ from its resonating frequency ω_0 can be expressed as

$$\mathcal{L}(\Delta\omega) = 10\log_{10}\left(\frac{KT}{2\,Q_t^2\,\alpha_I\,\alpha_V\,P_{DC}}\cdot F\cdot\left(\frac{\omega_0}{\Delta\omega}\right)^2\right), \qquad (6.2)$$

where K is the Boltzmann's constant; T is the absolute temperature; Q_t is the LC-tank quality factor; α_I is the current efficiency, defined as ratio of the fundamental current harmonic I_{ω_0} over the oscillator DC current I_{DC}; and α_V is the voltage efficiency, defined as a ratio of the single-ended oscillation amplitude, $V_{osc}/2$, over the supply voltage V_{DD} [21, 22]. Furthermore, F is the effective noise factor of the oscillator.

Equation (6.2) clearly demonstrates a trade-off between the oscillator's P_{DC} and PN. Furthermore, the oscillator's FoM normalizes the PN performance to the oscillation frequency and power consumption, yielding

$$FoM = 10\log_{10}\left(\frac{10^3 KT}{2\,Q_t^2\,\alpha_I\,\alpha_V}\cdot F\right). \qquad (6.3)$$

The effective noise factor F is expressed by [23, 24]

$$F = \frac{R_{in}}{2KT}\cdot\sum_i\frac{1}{2\pi}\int_0^{2\pi}\overline{i_{n,i}^2(\phi)}\cdot\Gamma_i^2(\phi)\,d\phi, \qquad (6.4)$$

where $\phi = \omega_0 t$, $\overline{i_{n,i}^2(\phi)}$ is the white current noise power density of the ith noise source and Γ_i is its relevant ISF function from the corresponding ith device noise [25]. Finally, R_{in} is an equivalent differential input parallel resistance of the tank's losses. The oscillator I_{DC} may be estimated by one of the following equations:

$$I_{DC} = \frac{I_{\omega_0}}{\alpha_I}\xrightarrow{I_{\omega_0}=\frac{V_{osc}}{R_{in}}}I_{DC} = \frac{V_{osc}}{R_{in}}\cdot\frac{1}{\alpha_I}\xrightarrow{V_{osc}=2\alpha_V V_{DD}}I_{DC} = \frac{2V_{DD}}{R_{in}}\cdot\frac{\alpha_V}{\alpha_I}. \qquad (6.5)$$

As a result, the RF oscillator's P_{DC} is derived by

$$P_{DC} = \frac{V_{DD}^2}{R_{in}}\cdot\frac{\alpha_V}{\alpha_I}. \qquad (6.6)$$

Equation (6.6) indicates that the minimum achievable P_{DC} can be expressed in terms of a set of *optimization* parameters, such as R_{in}, and a set of *topology*-dependent parameters, such as minimum supply voltage ($V_{DD,min}$), current and voltage efficiencies.

Lower P_{DC} is typically achieved by scaling up $R_{in} = L_p\omega_0 Q_t$ simply via a large multi-turn inductor, as in [26]. For example, while maintaining a constant Q_t, doubling L_P would theoretically double R_{in}, which would reduce P_{DC} by half but at a cost of a 3-dB PN degradation. However, at some point, that trade-off stops due to a dramatic drop in the inductor's self-resonant frequency and Q-factor. Figure 6.3(a) shows the simulated Q-factor of several multi-turn inductors in 28-nm CMOS versus their inductance values. As the inductor enlarges, the magnetic and capacitive coupling to the low-resistivity substrate increases such that the tank Q-factor drops almost linearly with L_P. As can be gathered from Figure 6.3(b), this constraint sets an upper limit on maximum $R_{in} = L_p\omega_0 Q_t$, which is chiefly a function of the technology node. Note that the inductor's value is largely dependent on its physical dimensions, rather than on the technology. However, the tank Q-factor is a bit degraded in the most recent process nodes (i.e., 28 nm) mainly due to more stringent minimum metal density rules, closer separation between the top-metal and substrate, as well as thinner lower-level metals that are used in metal-oxide-metal (MoM) capacitors. As a consequence, it is expected that $R_{in(max)}$ slightly reduces by migrating to finer CMOS technologies.

Parasitic capacitance of inductor windings, gm-devices, switchable capacitors, and oscillator routings determines a minimum floor of the tank's capacitance, which appears to be $\sim$250 fF at $f_0 = 4.8$ GHz. It puts another restriction on L_p and $R_{in(max)}$ to $\sim$4.5 nH and $\sim$1.3 kΩ and sets a lower limit on P_{DC} of each oscillator structure. Under this condition, the tank's Q-factor drops to $\leq$9. This explains the poor FoM of RF oscillators in modern BLE transceivers.

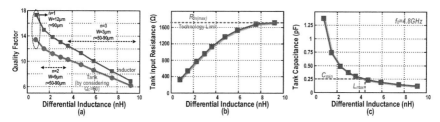

Figure 6.3 Dependency of various inductor parameters in 28-nm LP CMOS across inductance value: (a) inductor and tank Q-factor; (b) equivalent differential input resistance of the tank; and (c) required tank capacitance at 4.8-GHz resonance. Note that at this point the inductors are without dummy metal fills.

The topology-dependent parameters also play an important role in trying to reduce P_{DC}. Equation (6.6) favors structures that offer higher α_I or can sustain oscillation with smaller V_{DD} and α_V. On the other hand, $\alpha_V \cdot \alpha_I$ should be maximized to avoid any penalty on FoM [22,27], as evident from (6.2). Consequently, to efficiently reduce P_{DC} without disproportionately worsening the FoM, it is desired to employ structures with a higher α_I and a lower minimum V_{DD}. To get a better insight, Figure 6.4 shows such effects for the traditional cross-coupled NMOS-only (OSC$_N$) and complementary push–pull (OSC$_{NP}$) structures [28, 29]. Due to the less stacking of transistors, the $V_{DD,min}$ of OSC$_N$ can go 40% lower than that of OSC$_{NP}$. However, α_I of OSC$_{NP}$ is doubled due to the switching of tank current direction every half period. Its oscillation swing, and thus α_V, is also 50% smaller. Hence, OSC$_{NP}$ offers ~3× lower α_V/α_I. However, both structures demonstrate similar $\alpha_V \cdot \alpha_I$ product [30]. Consequently, each of them has its own set of advantages and drawbacks such that the minimum achievable P_{DC} and FoM is almost identical, as shown in Table 6.2. Note that applying a tail filtering technique to a class-B oscillator increases its α_V [22, 31], which is in line with the FoM optimization but against the P_{DC} reduction, as evident from (6.2) and (6.6). Furthermore, while maintaining the same R_{in}, a class-F$_3$ operation does not reduce P_{DC} of traditional oscillators since its minimum V_{DD}, α_V and α_I are identical to OSC$_N$ (see Chapter 3).

A push–pull class-C oscillator appears as an excellent choice for ULP applications due to its largest α_I and smallest α_V [32], as per Table 6.2. However, it needs an additional complex biasing circuitry (e.g., an opamp)

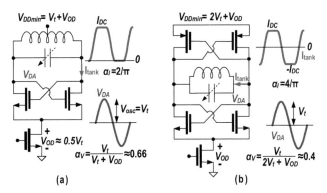

Figure 6.4 $V_{DD,min}$, α_I and α_V parameters for: (a) cross-coupled NMOS and (b) complementary push–pull oscillators.

Table 6.2 Minimum P$_{DC}$ for different RF oscillator topologies

Topology	$V_{DD,min}$[†]	α_V[‡]	α_I*	P_{DCmin}	$\alpha_V.\alpha_I$
OSC$_N$	$V_t + V_{OD} \approx 1.5\ V_t$	0.66	$2/\pi$	$4.66\ V_t^2/R_{in}$	0.42
OSC$_{NP}$	$2\ V_t + V_{OD} \approx 2.5\ V_t$	0.4	$4/\pi$	$3.92\ V_t^2/R_{in}$	0.51
OSC$_{NP}$ with tail filter	$2\ V_t + V_{OD} \approx 2.5\ V_t$	0.63	$4/\pi$	$6.2\ V_t^2/R_{in}$	0.8
Class-C$_{NP}$	$2\ V_t + V_{OD} \approx 2.5\ V_t$	0.25	2	0.15 mW $+ 1.56\ V_t^2/R_{in}$	0.5
Class-D	$\approx V_t$	1.635	0.5	$6.54\ V_t^2/R_{in}$	0.82
Class-F$_3$	$V_t + V_{OD} \approx 1.5\ V_t$	0.66	$2/\pi$	$4.7\ V_t^2/R_{in}$	0.42
This work	$V_t + V_{OD} \approx 1.5\ V_t$	0.33	$4/\pi$	$1.2\ V_t^2/R_{in}$	0.42

† by considering $V_{OD} = 0.5\ V_t$ for the current source.
‡ at the minimum V_{DD}.
* ideal value.

to guarantee the proper oscillator start-up and to keep the transistors in saturation during the on-state. There are also strong mutual trade-offs between the
biasing circuit's P_{DC}, oscillator's amplitude stability and PN, much intensified in ULP applications where the tank capacitance tends to be smaller [33].
As a consequence, the biasing circuitry can end up consuming comparable
power as the ULP oscillator itself. On the other hand, V_{DD} of class-D oscillators can go below a threshold voltage, V_t. However, due to hard switching of
core transistors, its α_V and α_I are, respectively, higher and lower than other
structures [34], as shown in Table 6.2. According to (6.6), this trend is against
the P_{DC} reduction. Consequently, the current oscillator structures have issues
with reaching simultaneous ultra-low power *and* voltage operation.

In this chapter, we disclose how to convert the fixed current source
of the traditional NMOS topology into a structure with alternating current
sources such that the tank current direction can change every half period.
Consequently, the benefits of low supply of the OSC$_N$ topology and higher
α_I of OSC$_{NP}$ structure are combined to reduce power consumption further
than practically possible in the traditional oscillators.

6.3 Switching Current-Source Oscillator

Figure 6.5 shows an evolution towards the switching current-source oscillator. The OSC$_N$ topology is chosen as a starting point due to its low V_{DD}
capability. To reduce P_{DC} further, it is desired to switch the direction of the

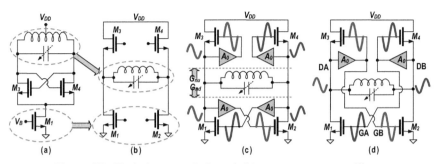

Figure 6.5   Evolution towards the switching current-source oscillator.

LC-tank current in each half period, which will double α_I. Consequently, we beneficially split the fixed current source M_1 in Figure 6.5(a) into two switchable current sources, M_1 and M_2, as suggested in Figure 6.5(b). This allows for the tank to be disconnected from the V_{DD} feed and be moved inbetween the upper and lower NMOS transistor pairs to give rise to an H-bridge configuration. In the next step, the passive voltage gain blocks, A_0, are added to the NMOS gates, as shown in Figure 6.5(c). Both upper and lower NMOS pairs should each independently demonstrate *synchronized* positive feedback to realize the switching of the tank current direction. The "master" positive feedback enforces the differential-mode operation and is realized by the lower-pair transistors configured in a conventional cross-coupled manner. Its negative conductance seen by the tank may be estimated by (see Section 6.4 for detailed calculations)

$$G_{down} = \frac{-A_0}{4} \cdot (g_{m1}(\phi) + g_{m2}(\phi)).$$ (6.7)

On the other upper side, the differential-mode oscillation of the tank is reinforced by the $M_{3,4}$ devices which realize the second positive feedback.[2] The negative conductance seen by the tank into the upper pair can be calculated by (see Section 6.4 for detailed calculations)

$$G_{up} = G_{down} = \frac{-(A_0 - 1)}{4} \cdot (g_{m3}(\phi) + g_{m4}(\phi)).$$ (6.8)

Equation (6.8) clearly indicates that the voltage gain block is necessary and A_0 must be safely larger than 1 to be able to present a negative conductance to the tank, thus enabling the H-bridge switching. By merging

---

[2]It should be noted that the "master/slave" view is mainly valid from a small-signal standpoint. Both are equally important when considering the large-signal switching operation.

the redundant voltage gain blocks, the disclosed switching current-source oscillator is shown in Figure 6.5(d). Note that the tank with an implicit voltage gain can be realized by using a capacitive divider, autotransformer, or step-up transformer, as illustrated in Figure 6.6. The transformer-based tank is chosen in this work due to its simplicity.

Figures 6.7 and 6.8 illustrate the novel oscillator schematic as well as waveforms and various operational regions of M_{1-4} transistors across the oscillation period. The two-port resonator consists of a step-up 1:2 transformer and tuning capacitors, C_1, C_2, at its primary and secondary windings.

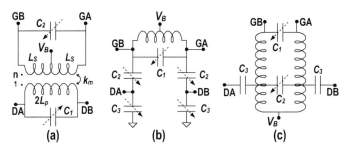

Figure 6.6   Various options of a tank providing voltage gain.

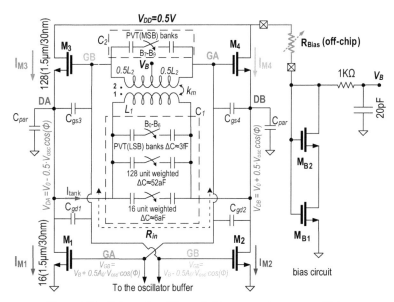

Figure 6.7   Schematic of the switching current-source oscillator.

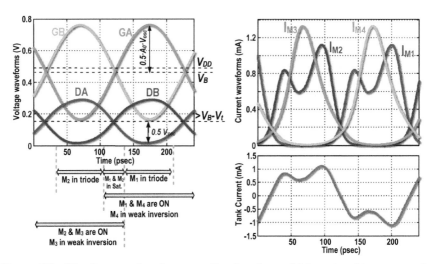

Figure 6.8 Waveforms and various operational regions of M_{1-4} transistors across the oscillation period.

The transistors $M_{1,2}$ set the oscillator's DC current, while the $M_{3,4}$ pair acts as a switching current source. Both $M_{1,2}$ and M_{3-4} pairs play an equally vital role of switching the tank current direction. As can be gathered from Figure 6.8, G_B oscillation voltage is high within the first half period. Hence, only M_2 and M_3 transistors are on and the current flows from left to right side of the tank. However, M_1 and M_4 are turned on for the second half period and the tank's current direction is reversed. Consequently, just like in the push–pull structure, the tank current flow is reversed every half period, thus doubling the oscillator's α_I to $4/\pi$.

The V_{DD} of the new oscillator can be as low as $V_{OD1} + V_{OD3} \approx V_t$, which is extremely small for an oscillator with a capability of switching the tank current direction. This makes it suitable for a direct connection to solar cells. Note that the oscillation swing cannot go further than $V_{OD1,2}$ at DA/DB nodes, which is chosen $\sim$150 mV to satisfy the system's phase noise requirement by a few dB margin. However, the maximum required voltage of the circuit is determined by the bias voltage V_B.

$$V_B \approx V_{OD1} + V_{gs3}. \tag{6.9}$$

Equation (6.9) implies that $M_{3,4}$ should work in weak inversion keeping $V_{gs3} < V_t$ to achieve lower $V_{DD,min}$. However, the transistor's cut-off frequency f_{max} drops dramatically in the subthreshold operation. Note that f_{max} should be at least $4\times$ higher than the operating frequency $f_0 = 4.8$ GHz

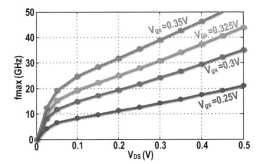

Figure 6.9 f_{max} of low-V_t 28 nm transistor versus V_{DS} for different V_{GS}.

to guarantee the oscillator start-up over PVT variations. This constraint limits $V_{gs3} \approx 0.32$ V for $V_{OD3} \approx 150$ mV, as can be gathered from Figure 6.9. Consequently, even by considering only the tougher V_B requirement, the new structure can operate at V_{DD} as low as 0.5 V, on par with OSC_N.

Such a low V_{DD} and oscillation swing can easily lead to start-up problems in the traditional structures. It will certainly increase power consumption, P_{buf}, of the following buffer, which would require more gain in order to provide a rail-to-rail swing at its output that is interpreted as a local oscillator (LO) clock. Fortunately, the transformer gain enhances the oscillation swing at $M_{1,2}$ gates to even beyond V_{DD}, guaranteeing the oscillator start-up and reduction of P_{buf}. Consequently, the oscillator buffer is connected to the secondary winding of the transformer in this design.

As can be gathered from Figure 6.8, the $M_{3,4}$ switching current-source transistors operate in a class-C manner as in a Colpitts oscillator, meaning that they deliver more or less narrow-and-tall current pulses. However, their non-zero conduction angle is quite wide, $\sim\pi$, due to the low overdrive voltage in the subthreshold operation. On the other hand, $M_{1,2}$ operate in a class-B manner like cross-coupled oscillators, meaning that they deliver square-shaped current pulses. Hence, the shapes of drain currents are quite different for the lower and upper pairs. However, their fundamental components demonstrate the same amplitude ($\alpha_I \approx 2/\pi$) and phase to realize the constructive oscillation voltage across the tank. The higher drain harmonics obviously show different characteristics. However, they are filtered out by the tank's selectivity characteristic. Note that the current through a transistor of the upper pair will have two paths to ground: through the corresponding transistor of the lower pair and through the single-ended capacitors. Consequently, the single-ended capacitors sink the higher current harmonics of $M_{3,4}$ transistors.

6.4 Thermal Noise Upconversion

To calculate a closed-form PN equation, the oscillator model is simplified in Figure 6.10. At the resonant frequency, the transformer-based tank can be modeled by an equivalent LC tank of elements L_{eq}, C_{eq}, and R_{in}.[3] On the other hand, M_{1-4} transistors with passive voltage gain of the transformer are decomposed into two nonlinear time-variant conductances. Note that the active elements in the circuit may add to the resonator loss, particularly at the extremes of large oscillation waveforms which may push transistors into their triode regions. Consequently, the nonlinearity is decomposed into two nonlinear resistances: one that is always positive, $G_{ds}(\phi)$, and one that is always negative, $G_n(\phi)$, where $\phi = \omega_0 t$. Further, to get a better insight, the effects of noise on the oscillator phase noise due to channel conductance $(\overline{i^2_{n,Gds}(\phi)} = 4KTG_{ds}(\phi))$ and transconductance gain $(\overline{i^2_{n,Gm}(\phi)})$ of M_{1-4} transistors are separately modeled in Figure 6.10. All circuit variables in this generic model will be obtained in the following sections.

6.4.1 Calculating the Effective Noise Due to Transconductance Gain of M_{1-4} Transistors $(\overline{i^2_{n,Gm}(\phi)})$

It is clear that the lower pair is a voltage-biased circuit. Consequently, the noise sources of M_1 and M_2 are uncorrelated for the entire oscillation period. However, the situation is more complicated for the upper pair. For a short time around zero-crossings, both transistors of the upper pair work in the sub-threshold region, while elsewhere one of them is off and the other device will be driven into saturation. In this situation, current through the upper NMOS transistor will have no path to ground other than through the corresponding

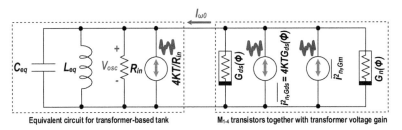

Equivalent circuit for transformer-based tank M_{1-4} transistors together with transformer voltage gain

Figure 6.10   Generic noise circuit model of the disclosed oscillator.

---

[3]The interested reader is directed to [35] for accurate closed-form equations of L_{eq}, C_{eq}, and R_{in}.

NMOS transistor of the lower pair if there were no single-ended capacitance at the tank. This phenomenon creates a common-mode oscillation across the tank, which ensures that the drain currents of both lower and upper NMOS transistor are the same.

However, if the tank includes some single-ended capacitors connected to ground, the oscillator will behave very differently (we also use this single-ended capacitors to create a common-mode resonant frequency at the second harmonic of the fundamental frequency to reduce 1/f noise upconversion). Note that one cannot avoid the presence of single-ended capacitors in the tank due to drain–bulk and drain–source parasitic capacitance of lower-pair transistors, source–bulk and drain–source parasitic capacitance of upper pair transistors, and parasitic capacitance of the transformer's primary winding. In this situation, the current through transistors of upper pair will have two paths to the ground: through the corresponding NMOS transistor of the lower pair and through the single-ended capacitors. Consequently, single-ended capacitors suppress the common-mode oscillation voltage across the tank. In this instance, the upper pair is more appropriately viewed as a voltage-biased circuit. Consequently, noise sources due to transconductance gain of M_{1-4} transistors are absolutely uncorrelated.

By using the same approach as [36], we are going to replace all noise sources with an equivalent noise source across the tank. By writing KCL at DA and DB nodes, it is straightforward to show that the instantaneous equivalent current can be calculated by (see Figure 6.11)

$$\left.\begin{array}{l} I_{eq} = I_{M1} - I_{M3} \\ I_{eq} = I_{M4} - I_{M2}, \end{array}\right\} \rightarrow I_{eq} = \frac{1}{2}((I_{M1} + I_{M4}) - (I_{M1} + I_{M4})) \rightarrow$$

$$I_{eq} = (I_{M1} - I_{M2}) + (I_{M3} - I_{M4}). \qquad (6.10)$$

As a consequence, the resulting differential noise current through the tank is

$$\overline{i_{n,Gm}^2}(\phi) = \frac{1}{4}\left(\overline{i_{n,gm1}^2}(\phi) + \overline{i_{n,gm2}^2}(\phi) + \overline{i_{n,gm3}^2}(\phi) + \overline{i_{n,gm4}^2}(\phi)\right) \rightarrow$$

$$\overline{i_{n,Gm}^2}(\phi) = KT\left(\gamma_1(g_{m1}(\phi) + g_{m2}(\phi)) + \gamma_3(g_{m3}(\phi) + g_{m4}(\phi))\right). \quad (6.11)$$

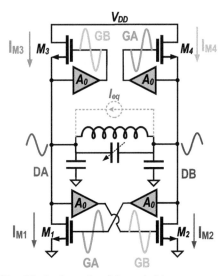

Figure 6.11   Simplified schematic of the switching current-source oscillator.

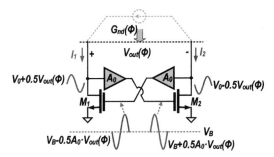

Figure 6.12   Simplified schematic of the lower pair of the oscillator.

6.4.2 Calculating the Negative Conductance of the Oscillator ($G_n(\phi)$)

The negative conductance of lower and upper pairs will be calculated separately in the following sections. The upper and lower negative conductances are in parallel. Hence, the total negative conductance is calculated by adding the negative conductance of lower and upper pairs.

The gate–source voltage of M_1 is calculated by (see Figure 6.12)

$$V_{GS1}(\phi) = V_B - \frac{A_0 V_{out}(\phi)}{2}. \tag{6.12}$$

As a result, the derivative of gate–source voltage of M_1 is calculated by

$$\frac{dV_{GS1}(\phi)}{d\phi} = -\frac{A_0}{2} \cdot \frac{dV_{out}(\phi)}{d\phi}. \tag{6.13}$$

The transconductance gain of M_1 transistor may be estimated by

$$g_{m1}(\phi) = \frac{dI_1(\phi)}{dV_{GS1}} = \frac{dI_1(\phi)/d\phi}{dV_{GS1}/d\phi} = \frac{dI_1(\phi)/d\phi}{-\frac{A_0}{2} dV_{out}/d\phi} = -\frac{2}{A_0} \cdot \frac{dI_1(\phi)}{dV_{out}(\phi)}. \tag{6.14}$$

We can rewrite Equation (6.14) as

$$\frac{dI_1(\phi)}{dV_{out}(\phi)} = -\frac{2}{A_0} \cdot g_{m1}(\phi). \tag{6.15}$$

On the other hand, the gate–source voltage of M_2 is calculated by

$$g_{m2}(\phi) = \frac{dI_2(\phi)}{dV_{GS2}} = \frac{dI_2(\phi)/d\phi}{dV_{GS2}/d\phi} = \frac{dI_2(\phi)/d\phi}{\frac{A_0}{2} dV_{out}/d\phi} = \frac{2}{A_0} \cdot \frac{dI_2(\phi)}{dV_{out}(\phi)}. \tag{6.16}$$

And again, we can rewrite Equation (6.16) as

$$\frac{dI_2(\phi)}{dV_{out}(\phi)} = \frac{2}{A_0} \cdot g_{m2}(\phi). \tag{6.17}$$

The effective negative conductance of lower pair,

$$G_{nd}(\phi) = \frac{dI_{eq2}(\phi)}{dV_{out}(\phi)} = \frac{1}{2} \cdot \frac{dI_1(\phi) - dI_2(\phi)}{dV_{out}(\phi)}. \tag{6.18}$$

By using (6.15) and (6.17), the above equation can be rewritten by

$$G_{nd}(\phi) = -\frac{1}{4} \cdot A_0 \cdot (g_{m1}(\phi) + g_{m2}(\phi)). \tag{6.19}$$

The same calculations can be done for M_3 and M_4.

The gate–source voltage of M_3 is calculated by (see Figure 6.13)

$$V_{GS3}(\phi) = V_B - V_0 + 0.5(A_0 - 1)V_{out}(\phi). \tag{6.20}$$

As a result, the derivative of gate–source voltage of M_3 is calculated by

$$\frac{dV_{GS3}(\phi)}{d\phi} = 0.5(A_0 - 1) \cdot \frac{dV_{out}(\phi)}{d\phi}. \tag{6.21}$$

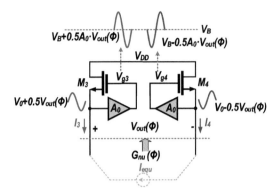

Figure 6.13 Simplified schematic of the upper pair of the oscillator.

The transconductance gain of M_3 transistor may be estimated by

$$g_{m3}(\phi) = \frac{dI_3(\phi)}{dV_{GS3}} = \frac{dI_3(\phi)/d\phi}{dV_{GS3}/d\phi} = \frac{dI_3(\phi)/d\phi}{0.5(A_0 - 1)\frac{dV_{out}}{d\phi}}$$

$$= \frac{2}{(A_0 - 1)} \cdot \frac{dI_3(\phi)}{dV_{out}(\phi)}. \tag{6.22}$$

We can rewrite the above equation by

$$\frac{dI_3(\phi)}{dV_{out}(\phi)} = \frac{A_0 - 1}{2} \cdot g_{m3}(\phi). \tag{6.23}$$

On the other hand, the gate–source voltage of M_4 is calculated by

$$V_{GS4}(\phi) = V_B - V_0 - \frac{(A_0 - 1)}{2} V_{out}(\phi). \tag{6.24}$$

As a result, the derivative of gate–source voltage of M_4 is calculated by

$$\frac{dV_{GS4}(\phi)}{d\phi} = \frac{(A_0 - 1)}{2} \cdot \frac{dV_{out}(\phi)}{d\phi}. \tag{6.25}$$

The transconductance gain of M_4 transistor then is estimated by

$$g_{m4}(\phi) = \frac{dI_4(\phi)}{dV_{GS4}} = \frac{dI_4(\phi)/d\phi}{dV_{GS4}/d\phi} = -\frac{dI_4(\phi)/d\phi}{0.5(A_0 - 1)\frac{dV_{out}}{d\phi}}$$

$$= -\frac{2}{(A_0 - 1)} \cdot \frac{dI_4(\phi)}{dV_{out}(\phi)}. \tag{6.26}$$

We can rewrite the above equation by

$$\frac{dI_4(\phi)}{dV_{out}(\phi)} = -\frac{A_0 - 1}{2} \cdot g_{m4}(\phi). \tag{6.27}$$

The negative conductance of upper pair

$$G_{nu}(\phi) = \frac{dI_{equ}(\phi)}{dV_{out}(\phi)} = \frac{1}{2} \cdot \frac{dI_4(\phi) - dI_3(\phi)}{dV_{out}(\phi)}. \tag{6.28}$$

By using (6.23) and (6.27), the above equation can be rewritten by

$$G_{nu}(\phi) = -\frac{1}{4} \cdot (A_0 - 1) \cdot (g_{m3}(\phi) + g_{m4}(\phi)). \tag{6.29}$$

The upper and lower negative conductance are in parallel. Hence, the total negative conductance is calculated by

$$G_n(\phi) = G_{nd}(\phi) + G_{nu}(\phi)$$
$$= \frac{1}{4} \cdot [A_0 \cdot (g_{m1}(\phi) + g_{m2}(\phi)) + (A_0 - 1) \cdot (g_{m3}(\phi) + g_{m4}(\phi))]. \tag{6.30}$$

6.4.3 Calculating the Positive Conductance of the Oscillator ($G_{DS}(\phi)$)

The positive conductance of lower and upper pairs will be calculated separately in the following sections. The upper and lower positive conductance are in parallel. Hence, the total positive conductance is calculated by adding the positive conductance of lower and upper pairs.

The drain–source voltage of M_1 is calculated by (see Figure 6.12)

$$V_{ds1}(\phi) = V_0 + \frac{V_{out}(\phi)}{2}. \tag{6.31}$$

As a result, the derivative of drain–source voltage of M_1 is calculated by

$$\frac{dV_{ds1}(\phi)}{d\phi} = \frac{1}{2} \cdot \frac{dV_{out}(\phi)}{d\phi}. \tag{6.32}$$

The drain–source conductance of M_1 transistor may be estimated by

$$g_{ds1}(\phi) = \frac{dI_1(\phi)}{dV_{ds1}} = \frac{dI_1(\phi)/d\phi}{dV_{ds1}/d\phi} = -\frac{dI_1(\phi)/d\phi}{0.5\frac{dV_{out}}{d\phi}} = 2 \cdot \frac{dI_1(\phi)}{dV_{out}(\phi)}. \tag{6.33}$$

We can rewrite the above equation by

$$\frac{dI_1(\phi)}{dV_{out}(\phi)} = \frac{1}{2} \cdot g_{ds1}(\phi). \tag{6.34}$$

On the other hand, the drain–source voltage of M_2 is calculated by

$$V_{ds2}(\phi) = V_0 - \frac{V_{out}(\phi)}{2}. \tag{6.35}$$

As a result, the derivative of drain–source voltage of M_2 is calculated by

$$\frac{dV_{ds2}(\phi)}{d\phi} = -\frac{1}{2} \cdot \frac{dV_{out}(\phi)}{d\phi}. \tag{6.36}$$

The drain–source conductance gain of M_2 transistor may be estimated by

$$g_{ds2}(\phi) = \frac{dI_2(\phi)}{dV_{ds2}} = \frac{dI_2(\phi)/d\phi}{dV_{ds2}/d\phi} = -\frac{dI_2(\phi)/d\phi}{-0.5\frac{dV_{out}}{d\phi}} = -2 \cdot \frac{dI_2(\phi)}{dV_{out}(\phi)}. \tag{6.37}$$

We can rewrite the above equation as

$$\frac{dI_2(\phi)}{dV_{out}(\phi)} = -\frac{1}{2} \cdot g_{ds2}(\phi). \tag{6.38}$$

The positive conductance of lower pair then will be

$$G_{ds-down}(\phi) = \frac{1}{2} \cdot \frac{dI_1(\phi) - dI_2(\phi)}{dV_{out}(\phi)}. \tag{6.39}$$

Using (6.34) and (6.38), we can conclude:

$$G_{ds-down}(\phi) = \frac{1}{4} \left(g_{ds1}(\phi) + g_{ds2}(\phi) \right). \tag{6.40}$$

The drain–source voltage of M_3 is calculated by (see Figure 6.13)

$$V_{ds3}(\phi) = V_{DD} - V_0 - \frac{V_{out}(\phi)}{2}. \tag{6.41}$$

As a result,

$$\frac{dV_{ds3}(\phi)}{d\phi} = -\frac{1}{2} \cdot \frac{dV_{out}(\phi)}{d\phi}. \tag{6.42}$$

The drain–source conductance gain of M_3 transistor then is estimated by

$$g_{ds3}(\phi) = \frac{dI_3(\phi)}{dV_{ds3}} = \frac{dI_3(\phi)/d\phi}{dV_{ds3}/d\phi} = \frac{dI_3(\phi)/d\phi}{-0.5\frac{dV_{out}}{d\phi}} = -2 \cdot \frac{dI_3(\phi)}{dV_{out}(\phi)}. \quad (6.43)$$

We can rewrite the above equation as

$$\frac{dI_3(\phi)}{dV_{out}(\phi)} = -\frac{1}{2} \cdot g_{ds3}(\phi). \quad (6.44)$$

On the other hand, the drain–source voltage of M_4 is calculated by

$$V_{ds4}(\phi) = V_{DD} - V_0 + \frac{V_{out}(\phi)}{2}. \quad (6.45)$$

As a result,

$$\frac{dV_{ds4}(\phi)}{d\phi} = +\frac{1}{2} \cdot \frac{dV_{out}(\phi)}{d\phi} \quad (6.46)$$

and

$$g_{ds4}(\phi) = \frac{dI_4(\phi)}{dV_{ds4}} = \frac{dI_4(\phi)/d\phi}{dV_{ds43}/d\phi} = \frac{dI_4(\phi)/d\phi}{0.5\frac{dV_{out}}{d\phi}} = 2 \cdot \frac{dI_4(\phi)}{dV_{out}(\phi)}. \quad (6.47)$$

We can rewrite the above equation as

$$\frac{dI_4(\phi)}{dV_{out}(\phi)} = \frac{1}{2} \cdot g_{ds4}(\phi). \quad (6.48)$$

The positive conductance of upper pair then will be

$$G_{ds-up}(\phi) = \frac{1}{2} \cdot \frac{dI_4(\phi) - dI_3(\phi)}{dV_{out}(\phi)}. \quad (6.49)$$

Using (6.44) and (6.48), we can conclude

$$G_{ds-up}(\phi) = \frac{1}{4} \left(g_{ds3}(\phi) + g_{ds4}(\phi) \right). \quad (6.50)$$

The upper and lower positive conductance are in parallel. Hence, the total negative conductance is calculated by

$$G_{ds}(\phi) = G_{ds-up}(\phi) + G_{ds-down} = \frac{1}{4} \left[g_{ds1} + g_{ds2} + g_{ds3}(\phi) + g_{ds4}(\phi) \right]. \quad (6.51)$$

6.4.4 Satisfying Barkhausen Criterion

To sustain oscillation, the average power dissipated in the tank (R_{in}) and positive conductance of active devices ($G_{ds}(\phi)$) must equal the average power delivered by the negative conductance of nonlinearity ($G_n(\phi)$) [36]. Hence,

$$P_{R_{in}} + P_{G_{ds}} = -P_{G_n}. \tag{6.52}$$

Assuming $V_{out} = A_c cos(\omega_0 t)$, the average power dissipated in the tank can be calculated by

$$P_{R_{in}} = \frac{1}{T} \int_{-\frac{T}{2}}^{\frac{T}{2}} \frac{(A_c cos(\omega_0 t))^2}{R_p} dt = \frac{A_c^2}{2R_p}. \tag{6.53}$$

The current drawn by the positive conductance of nonlinearity can be described as

$$I_{G_{ds}}(t) = I_{G_{ds}-DC} + \int_{-\infty}^{t} d(I_{G_{ds}}(\tau))d\tau$$

$$= I_{G_{ds}-DC} + \int_{-\infty}^{t} G_{ds}(\tau)d(V_{out}(\tau))d\tau$$

$$= I_{G_{ds}-DC} - A_c\omega_0 \int_{-\infty}^{t} G_{ds}(\tau) \cdot sin(\omega_0\tau))d\tau \tag{6.54}$$

and the average power dissipated by the positive conductance of the nonlinearity is

$$P_{G_{ds}}(t) = \frac{1}{T} \int_{-T/2}^{T/2} V_{out}(t) \cdot I_{G_{DS}}(t)dt$$

$$= \frac{1}{T} \int_{-T/2}^{T/2} A_c cos(\omega_0 t) \cdot \left[I_{G_{DS}-DC} - A_c\omega_0 \right.$$

$$\left. \int_{-\infty}^{t} G_{ds}(\tau) \cdot sin(\omega_0\tau) \cdot d\tau \right] \cdot dt$$

$$= \frac{1}{T} \int_{-T/2}^{T/2} A_c cos(\omega_0 t) \cdot I_{G_{DS}-DC} \cdot dt$$

$$= -\frac{A_c^2\omega_0}{T} \int_{-T/2}^{T/2} cos(\omega_0 t) \left[\int_{-\infty}^{t} G_{ds}(\tau) \cdot sin(\omega_0\tau)d\tau \right] \cdot dt. \tag{6.55}$$

If we switch the order of the integrals, we may write

$$P_{G_{ds}}(t) = -\frac{A_c^2 \omega_0}{T} \int_{-\frac{T}{2}}^{\frac{T}{2}} \int_{\tau}^{\frac{T}{2}} G_{ds}(\tau) \cdot sin(\omega_0 \tau) \cdot cos(\omega_0 t) \cdot dt \cdot d\tau$$

$$= -\frac{A_c^2}{T} \int_{-\frac{T}{2}}^{\frac{T}{2}} G_{ds}(\tau) \cdot sin(\omega_0 \tau) \cdot d\tau \cdot sin(\omega_0 t)|_{\tau}^{\frac{T}{2}}$$

$$= +\frac{A_c^2}{T} \int_{-\frac{T}{2}}^{\frac{T}{2}} G_{ds}(\tau) \cdot (sin(\omega_0 \tau))^2 \cdot d\tau$$

$$= +\frac{A_c^2}{T} \int_{-\frac{T}{2}}^{\frac{T}{2}} G_{ds}(\tau) \cdot (1 - cos(2\omega_0 \tau)) \cdot d\tau. \tag{6.56}$$

We know

$$G_{ds}[0] = \frac{1}{T} \int_{-T/2}^{T/2} G_{DS}(\tau) \cdot d\tau = \frac{1}{2\pi} \int_{-\pi}^{\pi} G_{DS}(\phi) \cdot d\phi \tag{6.57}$$

and

$$G_{ds}[2] = \frac{1}{T} \int_{-T/2}^{T/2} G_{DS}(\tau) \cdot cos(2\omega_0 \tau) \cdot d\tau$$

$$= \frac{1}{2\pi} \int_{-\pi}^{\pi} G_{DS}(\phi) \cdot cos(2\phi) \cdot d\phi, \tag{6.58}$$

where $G_{ds}[k]$ describes the Fourier series coefficients of the instantaneous positive conductance of nonlinearity $G_{ds}(t)$. By replacing (6.57) and (6.58) in (6.56)

$$P_{G_{ds}} = \frac{A_c^2}{2} \cdot (G_{DS}[0] - G_{DS}[2]). \tag{6.59}$$

We also define,

$$G_{DSEF} = G_{DS}[0] - G_{DS}[2]. \tag{6.60}$$

By replacing (6.60) in (6.59),

$$P_{G_{ds}} = \frac{A_c^2}{2} \cdot (G_{DSEF}). \tag{6.61}$$

Now let us calculate the average power delivered $G_n(\phi)$. The current drawn by the $G_n(\phi)$ can be described as

$$I_{G_n}(t) = I_{G_n-DC} + \int_{-\infty}^{t} d(I_{G_n}(\tau))d\tau$$

$$= I_{G_n-DC} + \int_{-\infty}^{t} G_n(\tau)d(V_{out}(\tau))d\tau$$

$$= I_{G_n-DC} - A_c\omega_0 \int_{-\infty}^{t} G_n(\tau) \cdot sin(\omega_0\tau))d\tau. \qquad (6.62)$$

and the average power dissipated by the positive conductance of the nonlinearity is

$$P_{G_n}(t) = \frac{1}{T} \int_{-T/2}^{T/2} V_{out}(t) \cdot I_{G_n}(t)dt$$

$$= \frac{1}{T} \int_{-T/2}^{T/2} A_c cos(\omega_0 t) \cdot \left[I_{G_{DS}-DC} \right.$$

$$\left. - A_c\omega_0 \int_{-\infty}^{t} G_n(\tau) \cdot sin(\omega_0\tau) \cdot d\tau \right] \cdot dt$$

$$= \frac{1}{T} \int_{-T/2}^{T/2} A_c cos(\omega_0 t) \cdot I_{G_n-DC} \cdot - \frac{A_c^2\omega_0}{T} \int_{-T/2}^{T/2} cos(\omega_0 t)$$

$$\left[\int_{-\infty}^{t} G_n(\tau) \cdot sin(\omega_0\tau)d\tau \right] \cdot dt. \qquad (6.63)$$

If we switch the order of the integrals, we may write

$$P_{G_n}(t) = -\frac{A_c^2\omega_0}{T} \int_{-\frac{T}{2}}^{\frac{T}{2}} \int_{\tau}^{\frac{T}{2}} G_n(\tau) \cdot sin(\omega_0\tau) \cdot cos(\omega_0 t) \cdot dt \cdot d\tau$$

$$= -\frac{A_c^2}{T} \int_{-\frac{T}{2}}^{\frac{T}{2}} G_n(\tau) \cdot sin(\omega_0\tau) \cdot d\tau \cdot sin(\omega_0 t)|_{\tau}^{\frac{T}{2}}$$

$$= +\frac{A_c^2}{T} \int_{-\frac{T}{2}}^{\frac{T}{2}} G_n(\tau) \cdot (sin(\omega_0\tau))^2 \cdot d\tau$$

$$= +\frac{A_c^2}{T} \int_{-\frac{T}{2}}^{\frac{T}{2}} G_n(\tau) \cdot (1 - cos(2\omega_0\tau)) \cdot d\tau. \qquad (6.64)$$

Consequently,

$$P_{G_n} = \frac{A_c^2}{2} \cdot (G_{NSEF}). \qquad (6.65)$$

To sustain oscillation, the average power dissipated in the tank (R_{in}) and positive conductance of active devices (G_{ds}) must equal the average power delivered by the negative conductance of nonlinearity (G_n). By replacing (6.53), (6.61) and (6.65) in (6.52),

$$\frac{A_c^2}{2R_{in}} + \frac{A_c^2}{2} \cdot (G_{DSEF}) = -\frac{A_c^2}{2} \cdot (G_{NEF}). \tag{6.66}$$

Consequently,

$$G_{NEF} = -\frac{1 + R_{in} \cdot G_{DSEF}}{R_{in}}. \tag{6.67}$$

On the other hand, the total effective negative conductance can be rewritten as sum of the effective negative conductance of lower and upper pairs

$$G_{NDEF} + G_{NUEF} = -\frac{1 + R_{in} \cdot G_{DSEF}}{R_{in}}. \tag{6.68}$$

Note that both upper and lower NMOS pairs should each individually demonstrate synchronized positive feedback to realize the switching of the tank current direction. Consequently, as with traditional complimentary oscillator, each pair should roughly compensate half of the oscillator losses.

$$G_{NDEF} = G_{NUEF} = -\frac{1}{2} \cdot \frac{1 + R_{in} \cdot G_{DSEF}}{R_{in}}. \tag{6.69}$$

On the other hand, G_{NDEF} and G_{NUEF} can be, respectively, calculated by

$$G_{NDEF} = -\frac{1}{4} \cdot A_0 \cdot [G_{M1EF} + G_{M2EF}] \tag{6.70}$$

and

$$G_{NUEF} = -\frac{1}{4} \cdot (A_0 - 1) \cdot [G_{M3EF} + G_{M4EF}]. \tag{6.71}$$

By merging (6.70), (6.71) and (6.69), we have

$$G_{M1EF} + G_{M2EF} = \frac{2}{(A_0)} \cdot \frac{1 + R_{in} \cdot G_{DSEF}}{R_{in}} \tag{6.72}$$

$$G_{M3EF} + G_{M4EF} = \frac{2}{(A_0 - 1)} \cdot \frac{1 + R_{in} \cdot G_{DSEF}}{R_{in}}. \tag{6.73}$$

Since the oscillator is a symmetric circuit, the effective transconductance of M_1 and M_2 (also, M_3 and M_4) are the same. Hence, we can rewrite the above equation by

$$G_{M1EF} = \frac{1}{(A_0)} \cdot \frac{1 + R_{in} \cdot G_{DSEF}}{R_{in}} \tag{6.74}$$

and

$$G_{M4EF} = \frac{1}{(A_0 - 1)} \cdot \frac{1 + R_{in} \cdot G_{DSEF}}{R_{in}}. \tag{6.75}$$

We will use (6.74) and (6.75) later for calculating a closed-form equation of this oscillator.

6.4.5 Phase Noise Equation

It is well known that the phase noise and FoM of any RF oscillator at an offset frequency ω_0 from its resonating frequency $\omega_0 = 2\pi f_0$ can be expressed by

$$L(\Delta\omega) = 10 \log_{10} \left(\frac{KT}{2 \cdot Q_t^2 \cdot \alpha_I \cdot \alpha_V \cdot P_{DC}} \cdot \left(\frac{\omega_0}{\Delta\omega}\right)^2 \right) \tag{6.76}$$

and,

$$FoM = 10 \log_{10} \left(\frac{10^3 \cdot K \cdot T}{2 \cdot Q_t^2 \cdot \alpha_I \cdot \alpha_V} \cdot F \right), \tag{6.77}$$

where K is the Boltzmann's constant; T is the absolute temperature; Q_t is the LC-tank quality factor; α_I is the current efficiency, defined as ratio of the fundamental current harmonic I_{ω_0} over the oscillator DC current I_{DC}; and α_V is the voltage efficiency, defined as ratio of single-ended oscillation amplitude $V_{osc}/2$ over the supply voltage V_{DD}. F is the oscillator's effective noise factor and estimated by

$$F = \frac{R_{in}}{2KT} \cdot \sum_k \frac{1}{2\pi} \int_0^{2\pi} \overline{i_{n,k}^2(\phi)} \cdot \Gamma_k^2(\phi) d\phi. \tag{6.78}$$

Let us now calculate the contribution of the losses and active devices. The white current noise power density of the resistive loss of the oscillator is given by

$$\overline{i_{n,loss}^2(\phi)} = \overline{i_{n,tank}^2(\phi)} + \overline{i_{n,G_{ds}}^2(\phi)} = 4KT \left(\frac{1}{R_{in}} + G_{ds}(\phi) \right). \tag{6.79}$$

The relevant impulse sensitivity function of noise sources associated with a sinusoidal waveform oscillator, $V_{osc} \cdot \cos(\phi)$, may be estimated by $\Gamma = \sin(\phi)$ [25, 28]. By exploiting (6.4), the effective noise factor due to resistive losses of the oscillator becomes

$$
\begin{aligned}
F_{loss} &= \frac{R_{in}}{2KT} \cdot \frac{1}{2\pi} \int_0^{2\pi} \overline{i^2_{n,loss}(\phi)} \Gamma^2_{loss}(\phi) d\phi \\
&= \frac{R_{in}}{2KT} \cdot \frac{1}{2\pi} \int_0^{2\pi} 4KT \left(\frac{1}{R_{in}} + G_{ds}(\phi) \right) \cdot \sin^2(\phi) \cdot d\phi \\
&= \frac{1}{2\pi} \int_0^{2\pi} 2 \sin^2(\phi) d\phi + R_{in} \left(\frac{1}{2\pi} \int_0^{2\pi} G_{ds}(\phi) \cdot d\phi \right. \\
&\quad \left. - \frac{1}{2\pi} \int_0^{2\pi} G_{ds}(\phi) \cdot \cos(2\phi) \cdot d\phi \right) \to F_{loss} \\
&= 1 + R_{in} \left(G_{DS}[0] - G_{DS}[2] \right) = 1 + R_{in} G_{DSEF}, \quad (6.80)
\end{aligned}
$$

where $G_{DS}[k]$ describes the kth Fourier coefficient of the instantaneous $G_{ds}(\phi)$. From (6.51) and since the oscillator is a symmetric circuit, $G_{DSEF} = \frac{1}{4} \cdot [G_{DSEF1} + G_{DSEF2} + G_{DSEF3} + G_{DSEF4}] = \frac{1}{2} \cdot [G_{DSEF1} + G_{DSEF4}]$. Consequently, we can rewrite (6.80) as

$$
F_{loss} = 1 + \frac{R_{in}}{2} \left(G_{DS1EF} + G_{DS4EF} \right). \quad (6.81)
$$

To get a better insight, different components of the above equation are graphically illustrated in Figure 6.14(a)–(c). The literature interprets $R_{in}G_{DSEF}$ term in (6.81) as the tank loading effect. In our design, M_1 and M_2 alternatively enter the triode region for part of the oscillation period and exhibit a large channel conductance. As shown in Figure 6.14(a), simulated $0.5R_{in}G_{DS1EF}$ can be as large as 0.6 for the lower-pair transistors. However, $M_{3,4}$ work only in saturation and demonstrate small channel conductance for their entire on-state operation, as evident from Figure 6.14(a). Hence, the simulated value of $0.5R_{in}G_{DS4EF}$ is as low as 0.15 for upper pair transistors. Note that both NMOS and PMOS pairs of the OSC$_{NP}$ structure simultaneously enter the triode region for part of the oscillation period and load the tank from both sides. In this structure, however, only one side of the tank is connected to the AC ground when either M_1/M_2 is in triode while the other side sees high impedance. Hence, this structure at least preserves the charge of differential capacitors over the entire oscillation period. Consequently, compared to the traditional oscillators, the tank loading effect is somewhat reduced here.

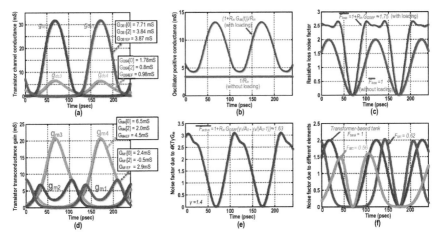

Figure 6.14 Circuit-to-phase-noise conversion across the oscillation period in the switching current-source oscillator. Simulated (a) channel conductance of M_{1-4}; (b) conductance due to resistive losses; (c) noise factor due to losses; (d) transconductance of M_{1-4}; (e) effective noise factor due to transconductance gain; (f) effective noise factors due to different oscillator's components.

The effective noise factor due to transconductance gain can be calculated by

$$F_{active} = \frac{R_{in}}{2KT} \cdot \frac{1}{2\pi} \int_0^{2\pi} \overline{i_{Gm}^2(\phi)} \Gamma_{Gm}^2(\phi) d\phi. \tag{6.82}$$

Replacing (6.11) in (6.82),

$$
\begin{aligned}
F_{active} &= \frac{R_{in}}{2KT} \cdot \frac{1}{2\pi} \int_0^{2\pi} KT\left(\gamma_1(g_{m1}(\phi) + g_{m2}(\phi)) + \gamma_3(g_{m3}(\phi)\right. \\
&\quad \left. + g_{m4}(\phi))\right) \cdot \sin^2(\phi) \cdot d\phi \\
&= \frac{R_{in}}{4\pi} \int_0^{2\pi} 2\sin^2(\phi)\left(\gamma_1(g_{m1}(\phi) + g_{m2}(\phi))\right. \\
&\quad \left. + \gamma_3(g_{m3}(\phi) + g_{m4}(\phi))\right) \cdot \left(\frac{1}{2} - \frac{1}{2}\cos(2\phi)\right) \\
&= \frac{R_{in}}{4}\left[\gamma_1(G_{M1}[0] - G_{M1}[2] + G_{M2}[0] - G_{M2}[2])\right. \\
&\quad \left. + \gamma_4(G_{M4}[0] - G_{M4}[2] + G_{M4}[0] - G_{M4}[2])\right] \\
&= \frac{R_{in}}{4}\left[\gamma_1(G_{M1EF} + G_{M2EF}) + \gamma_4(G_{M3EF} + G_{M4EF})\right] \tag{6.83}
\end{aligned}
$$

By replacing (6.72) and (6.73) in (6.83),

$$F_{active} = (1 + R_{in}G_{DSEF}) \cdot \left(\frac{\gamma_1}{2A_0} + \frac{\gamma_4}{2(A_0 - 1)} \right). \qquad (6.84)$$

To get a better insight, different components of the above equation are graphically illustrated in Figure 6.14(d–e).

As discussed in conjunction with Figure 6.5(c), the transformer's passive voltage gain, A_0, covers a significant part of the required loop gain of the lower positive feedback. Hence, the lower-pair transistors have to compensate only $1/(2A_0)$ of the circuit losses. For the upper positive feedback, however, A_0 covers a smaller part of the required loop gain. Consequently, the upper transistors should work harder and compensate $1/(2(A_0-1))$ of the oscillator loss. Consequently, as (6.84) indicates, the G_M noise contribution by the lower pair is smaller. However, its effect on F_{loss} is larger such that both pairs demonstrate more or less the same contribution to the oscillator PN (see Figure 6.14(f)). Finally, the total oscillator effective noise factor is given by

$$F = F_{loss} + F_{active} = (1 + R_{in}G_{DSEF}) \cdot \left(1 + \frac{\gamma_1}{2A_0} + \frac{\gamma_4}{2(A_0 - 1)} \right).$$
$$(6.85)$$

To obtain the oscillator phase noise, G_{DS1EF} and G_{DS4EF} should also be calculated or simulated. Since transistor size and oscillation waveforms are known, it is pretty straight-forward to calculate a closed-form equation for them. However, the final equation will be huge and these parameters are calculated numerically here.

When $M_{3,4}$ are not turned off, they work only in saturation and thus their channel conductance and G_{DS4EF} are negligible. However, as shown in the manuscript, precise simulations show that $\frac{R_{in}}{2} \cdot G_{DS4EF}$ can be as large as 0.15 even if the transistor works only in the saturation. It translates to 0.6 dB higher noise factor for this oscillator due to channel conductance noise of $M_{3,4}$ transistors. On the other hand, M_1 and M_2 alternatively enter the triode region for part of the oscillation period. Hence, their effective conductance G_{DS1EF} is larger. Simulations show that $\frac{R_{in}}{2} \cdot G_{DS1EF}$ is about 0.6 in this oscillator. We will also show later that the excess noise factor of NMOS transistors is 1.4. The voltage gain is 2.16. By replacing those numbers in the nose factor equation, we have

$$F = (1 + 0.6 + 0.15) \cdot \left[1 + \left(\frac{1.4}{2 \cdot 2.15} + \frac{1.4}{2 \cdot 1.15} \right) \right] \approx 5.3 \; dB \quad (6.86)$$

the noise factor is just 1.5 dB higher than the ideal value of $(1 + \gamma)$, despite the aforementioned practical issues of designing ulta-low voltage and power oscillators. The phase noise and FoM of this oscillator can be calculated by replacing (6.85) in (6.2).

6.5 1/f Noise Upconversion

Several techniques have been exploited to improve the oscillator's 1/f noise upconversion. First, dynamically switching the bias-setting devices $M_{1,2}$ will reduce their flicker noise, as also demonstrated in [37]. It also lowers a DC component of their effective impulse sensitivity function. Second, as discussed in Chapter 5, [38, 39], a second-order harmonic of the gm-devices' drain current flows into the capacitive part of the tank due to its lower impedance and creates asymmetric rise and fall times for the oscillation waveform. It directly increases a DC value of the oscillator ISF and thus its $1/f^3$ PN corner. This phenomenon can be alleviated by realizing an auxiliary resonance at $2\omega_0$ such that the second harmonic current flows into equivalent resistance of the tank in order to avoid disturbing the rise and fall time symmetry of the oscillation voltage. Since common-mode signals, such as a second harmonic of the drain current, cannot see the tuning capacitance at the transformer secondary winding [21], the auxiliary $2\omega_0$ can be realized without die area penalty and by adjusting the single-ended capacitance at the transformer primary winding [39].

The last source of the 1/f noise is M_{B1} in the biasing circuitry. By exploiting long channel device for biasing, its power consumption becomes negligible compared to the oscillator core while $M_{B1/B2}$ occupy larger area and thus generate lower 1/f noise. Consequently, based on aforementioned techniques, a lower $1/f^3$ PN corner is expected than in the traditional oscillators.

6.6 Optimizing Transformer-Based Tank

The transformer-based tank's input equivalent resistance, R_{in}, and voltage gain, A_0, should be maximized for the best system efficiency. Both optimization parameters are a strong function of $\zeta = L_2 C_2 / L_1 C_1$ [35], as shown in

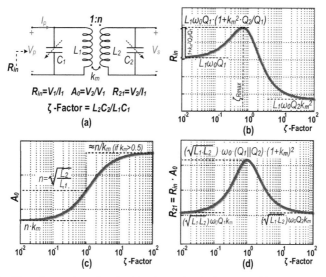

Figure 6.15 Transformer-based tank: (a) schematic; (b) input parallel resistance; (c) voltage gain; and (d) R_{21} versus ζ-factor.

Figure 6.15. R_{in} may be estimated by

$$R_{in} = L_1\omega_0 Q_1 \cdot \frac{\left(1 - \left(\frac{\omega_0}{\omega_s}\right)^2 (1 - k_m^2)\right)\zeta}{-\left(\frac{\omega_0}{\omega_s}\right)^4 \left(1 + \frac{Q_1}{Q_2}\right) + \left(\frac{\omega_0}{\omega_s}\right)^2 \left(1 + \frac{Q_1}{Q_2}\zeta\right)}, \quad (6.87)$$

where $\omega_s^2 = 1/L_2 C_2$, and Q_1 and Q_2 are, respectively, the Q-factors of the transformer's primary and secondary windings. It can be shown that R_{in} reaches its maximum when

$$\zeta_{Rmax} = \frac{Q_2}{Q_1} \cdot \left(\frac{Q_2}{Q_1 + Q_2} \cdot k_m^2 + \frac{Q_1}{Q_1 + Q_2}\right). \quad (6.88)$$

Note that the tank Q-factor is maximized at different $\zeta = Q_2/Q_1$ [24]. The maximum R_{in} is obtained by inserting (6.88) into (6.87)

$$R_{inmax} = L_1\omega_0 Q_1 \cdot \left(1 + k_m^2 \cdot \frac{Q_2}{Q_1}\right). \quad (6.89)$$

Consequently, the transformer's coupling factor k_m enhances R_{in} by a factor of $\sim(1 + k_m^2)$ at ζ_{Rmax}. For this reason, the switched-capacitor banks are distributed between the transformer's primary and secondary to roughly satisfy (6.88). For $k_m \geq 0.5$, the voltage gain of the transformer-based tank may be estimated by

$$A_0 = \frac{2k_m n}{1 - \zeta + \sqrt{1 + \zeta^2 + \zeta(4k_m^2 - 2)}}. \tag{6.90}$$

As shown in Figure 6.15(c), A_0 increases with larger ζ. Note that larger R_{in} and A_0 are desired to reduce P_{DC} and P_{buf}, respectively. To consider both scenarios, trans-impedance $R_{21} = R_{in} \cdot A_0$ term is defined and depicted in Figure 6.15(d). R_{21} reaches its maximum at $\zeta = 1$ for $Q_1 \approx Q_2$, which is reasonable for monolithic transformers. We also define the maximum of R_{21} as the transformer FoM $= (Q_1 \| Q_2) \cdot (1 + k_m)^2 \cdot \sqrt{L_1 L_2} \cdot \omega_0$. Consequently, the transformer dimensions and winding spacing are chosen to maximize this term.

6.7 Experimental Results

The oscillator was prototyped in TSMC 40 nm 1P7M CMOS. The chip micrograph is shown in Figure 6.16(a). $M_{1,2}$ and $M_{3,4}$ transistors are minimum-length low-V_t devices with a width of 32 and 256 μm, respectively. The transformer's primary and secondary differential self-inductance is only 660 pH and 2 nH, respectively, with the coupling factor $k_m = 0.76$. Both

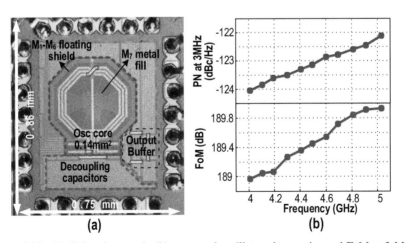

Figure 6.16   (a) Chip micrograph; (b) measured oscillator phase noise and FoM at 3-MHz offset frequency across the tuning range.

transformer's winding are realized with top ultra-thick metal (3.5 μm). However, the transformer includes a floating M1-to-M6 shield to comply with the strict metal density rules (>10%–20%) for manufacturability and also to alleviate the substrate loss. Note that the shield must be significantly thinner than the skin depth at the desired frequency to avoid any attenuation of the magnetic field. The skin depth of copper is ∼0.9 μm at 5 GHz. However, the thickness of M6 layer is 0.85 μm. Hence, adding M6 dummy metal reduces the transformer's magnetic field, inductance, and Q-factor, and thus R_{in} drops by 10%–20%. The simulated Q-factor is 12 and 16 for the primary and secondary windings, respectively.

Figure 6.17 shows the measured PN at the highest and lowest frequencies (f_{max}, f_{min}) with V_{DD} of 0.5 V and P_{DC} of 470 and 580 μW, respectively. Thanks to the switching current-source technique, $1/f^3$ PN corner of the oscillator is relatively low and varies between 250 and 420 kHz

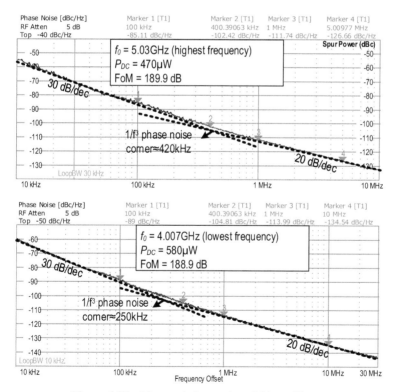

Figure 6.17   Measured phase noise of this oscillator.

Table 6.3   Comparison table of low power oscillators

	This Work	[40] JSSC'05	[23] JSSC'08	[41] ESS-CIRC'14[†]	[26] ISSCC'14
Technology	40 nm	0.18 μm	0.13 μm	28 nm	40 nm
V_{DD}	0.5 V	0.5 V	1 V	0.5 V	1 V
TR(%)	22.2	8.7	14	N/A	24.5
f_0(GHz)	4.8	3.8	4.9	2.35	2.44
PN (dBc/Hz)[‡]	−139	−143	−149.5	−125.8	−131.1
P_{DC} (mW)	0.48	0.57	1.4	0.38	0.4
FoM (dB)	189.8	193	195.5	187.5	183
FoM_T (dB)[*]	196.7	191.7	198.5	N/A	190.8
Freq pushing	17 MHz/V	273 MHz/V	N/A	N/A	N/A
Dummy fill	Yes	No	No	No	No
Area (mm²)	0.14	0.23	0.11	0.2	0.15
Oscillator topology	Switching current source	TRX feedback	Class-C	Class-D	Traditional

[†]Including LDO. LDO also performs a start-up role.
[‡]At $\Delta f = 10$ MHz normalized to 2.4-GHz carrier.
[*]$FOM_T = |PN|+20 \log_{10}((f_0/\Delta f)(TR/10)) - 10 \log_{10}(P_{DC}(mW))$.

across the tuning range (TR). The oscillator has a 22.2% TR, from 4 to 5 GHz. Figure 6.16(b) displays plots of phase noise and FoM across the TR. The FoM reaches maximum 189.9 dBc at f_{max} and varies ~1 dB across the TR.

Table 6.3 summarizes the oscillator performance and compares it with relevant state-of-the-art for P_{DC}<2 mW and TR>8%. It is the only one with the all-layer dummy metal fills inside the LC tank for manufacturability. For the similar P_{DC} (400–600 μW), only the transformer-feedback VCO [40] shows better FoM but with a much larger area, lower TR, and extremely high frequency pushing. Class-C VCO [23] also shows better FoM but at a much higher P_{DC}. Furthermore, it needs additional complex biasing circuits (such as opamp) for proper operation, which can potentially limit its minimum V_{DD} and thus P_{DC}.

It might be interesting to point out that switching current source oscillator is already adapted in a fractional-N ADPLL for BLE [42], in a fully integrated BLE transmitter [18], and a BLE transiver [43].

6.8 Conclusion

A switching current-source oscillator has been described and analyzed, providing deep insights into beneficial circuit operation. It combines advantages of low supply voltage operation of the conventional NMOS cross-coupled oscillator with high current efficiency of the complementary push–pull oscillator to reduce the oscillator supply voltage and dissipated power without sacrificing its start-up robustness or loading tank's Q-factor. The 28-nm CMOS prototype exhibits 189.5 dBc/Hz FoM, with 22% tuning range, dissipating 0.5 mW from 0.5 V power supply, while complying with the process technology manufacturing rules.

References

[1] Y.-H. L., C. Bachmann, X. Wang, Y. Zhang, A. Ba, B. Busze, M. Ding, P. Harpe, G.-J. van Schaik, G. Selimis, H. Giesen, J. Gloudemans, A. Sbai, L. Huang, H. Kato, G. Dolmans, K. Philips, and H. de Groot, "A 3.7 mW-RX 4.4 mW-TX fully integrated Bluetooth Low-Energy/IEEE802.15.4/proprietary SoC with an ADPLL-based fast frequency offset compensation in 40 nm CMOS," *IEEE International Solid-State Circuits Conference Digest of Technical Papers (ISSCC)*, Feb. 2015, pp. 236–237.

[2] T. Sano, M. Mizokami, H. Matsui, K. Ueda, K. Shibata, K. Toyota, T. Saitou, H. Sato, K. Yahagi, and Y. Hayash, "A 6.3 mW BLE transceiver embedded RX image-rejection filter and TX harmonic-suppression filter reusing on-chip matching network," *IEEE International Solid-State Circuits Conference Digest of Technical Papers (ISSCC)*, Feb. 2015, pp. 240–241.

[3] Prummel, M. Papamichail, J. Willms, R. Todi, W. Aartsen, W. Kruiskamp, J. Haanstra, E. Opbroek, S. Rievers, P. Seesink, J. van Gorsel, and H. Woering, "A 10 mW Bluetooth Low-Energy Transceiver With On-Chip Matching," *IEEE J. Solid-State Circuits*, vol. 50, no. 12, pp. 3077–3088, Dec. 2015.

[4] A. Wong, M. Dawkins, G. Devita, N. Kasparidis, A. Katsiamis, O. King, F. Lauria, J. Schiff, and A. Burdett, "A 1V 5 mA multimode IEEE 802.15.6/Bluetooth low-energy WBAN transceiver for biotelemetry applications," *IEEE International Solid-State Circuits Conference Digest of Technical Papers (ISSCC)*, 2012, pp. 300–301.

[5] G. Devita, A. Wong, M. Dawkins, K. Glaros, U. Kiani, F. Lauria, V. Madaka, O. Omeni, J. Schiff, A. Vasudevan, L. Whitaker, S. Yu, and

A. Burdett, "A 5mW multi-standard Bluetooth LE/IEEE 802.15.6 SoC for WBAN applications," *Proceedings of European Solid-state Circuits Conference (ESSCIRC)*, 2014, pp. 283–286.

[6] J. Masuch and M. Delgado-Restituto, "A 1.1-mW-RX 81.4 dBm sensitivity CMOS transceiver for Bluetooth Low Energy," *IEEE Transactions on Microwave Theory and Techniques*, vol. 61, no. 4, pp. 1660–1673, Apr. 2013.

[7] "CC2640 SimpleLinkTM Bluetooth smart wireless MCU," in *Texas Instruments*, 2015, pp. 1–57. Available: http://www.ti.com/lit/ds/symlink/cc2640.pdf.

[8] F.-W. Kuo, M. Babaie, R. Chen, K. Yen, J.-Y. Chien, L. Cho, F. Kuo, C.-P. Jou, F.-L. Hsueh, and R. Staszewski, "A fully integrated 28 nm Bluetooth Low-Energy transmitter with 36% system efficiency at 3 dBm," *Proceedings of European Solid-state Circuits Conference (ESSCIRC)*, 2015, pp. 356–359.

[9] A. Selvakumar, M. Zargham, and A. Liscidini, "Sub-mW current re-use receiver front-end for wireless sensor network applications," *IEEE J. Solid-State Circuits*, vol. 50, no. 12, pp. 2965–2974, Dec. 2015.

[10] F. Zhang, Y. Miyahara, and B. P. Otis, "Design of a 300-mV 2.4-GHz receiver using transformer-coupled techniques," *IEEE J. Solid-State Circuits*, vol. 48, no. 12, pp. 3190–3205, Dec. 2013.

[11] C. Bachmann, M. Vidojkovic, X. Huang, M. Lont, Y.-H. Liu, M. Ding, B. Busze, J. Gloudemans, H. Giesen, A. Sbai, G.-J. van Schaik, N. Kiyani, K. Kanda, K. Oishi, S. Masui, K. Philips, and H. de Groot, "A 3.5 mW 315/400 MHz IEEE802.15.6/proprietary mode digitally-tunable radio SoC with integrated digital baseband and MAC processor in 40nm CMOS," *Proceedings of IEEE VLSI Circuits Symposium*, 2015, pp. 94–95.

[12] P. Whatmough, G. Smart, S. Das, Y. Andreopoulos, and D. Bull, "A 0.6 V All-Digital body-coupled wakeup transceiver for IoT applications," *Proceedings of IEEE VLSI Circuits Symposium*, 2015, pp. 98–99.

[13] S. Bandyopadhyay and A. P. Chandrakasan, "Platform architecture for solar, thermal, and vibration energy combining with MPPT and single inductor," *IEEE J. Solid-State Circuits*, vol. 47, no. 9, pp. 2199–2215, Sept. 2012.

[14] K. Kadirvel, Y. Ramadass, U. Lyles, J. Carpenter, V. Ivanov, V. McNeil, A. Chandrakasan, and B. Lum-Shue-Chan, "A 330 nA

energy harvesting charger with battery management for solar and thermoelectric energy harvesting," *IEEE International Solid-State Circuits Conference Digest of Technical Papers (ISSCC)*, 2012, pp. 106–108.

[15] J. Kim, P. K. T. Mok, and C. Kim, "A 0.15V-input energy-harvesting charge pump with switching body biasing and adaptive dead-time for efficiency improvement," *IEEE International Solid-State Circuits Conference Digest of Technical Papers (ISSCC)*, 2014, pp. 394–395.

[16] X. Liu and E. Sanchez-Sinencio, "A 0.45-to-3V reconfigurable charge-pump energy harvester with two-dimensional MPPT for Internet of Things," *IEEE International Solid-State Circuits Conference Digest of Technical Papers (ISSCC)*, 2015, pp. 370–371.

[17] M. Babaie, M. Shahmohammadi, and R. B. Staszewski, "A 0.5 V 0.5 mW switching current source oscillator," *Proceedings of IEEE Radio Frequency Integrated Circuits (RFIC) Symposium*, 2015, pp. 356–359.

[18] M. Babaie, F. W. Kuo, H. N. Ron Chen, L. C. Cho, C. P. Jou, F. L. Hsueh, Mina Shahmohammadi, and Robert Bogdan Staszewski, "A fully integrated Bluetooth Low-Energy transmitter in 28 nm CMOS with 36% system efficiency at 3 dBm" *IEEE J. Solid-State Circuits*, vol. 51, no. 7, pp. 1547–1565, Jul. 2016.

[19] A. A. Abidi, "Phase noise and jitter in CMOS ring oscillators," *IEEE Journal of Solid-State Circuits,* vol. 41, no. 8, pp. 1803–1816, Aug. 2006.

[20] "Bluetooth Specification Version 4.2," in *Bluetooth*, 2014. Available: http://www.bluetooth.com.

[21] M. Babaie and R. B. Staszewski, "An ultra-low phase noise class-F2 CMOS oscillator with 191 dBc/Hz FoM and long-term reliability," *IEEE J. Solid-State Circuits*, vol. 50, no. 3, pp. 679–692, Mar. 2015.

[22] M. Garampazzi, S. Dal Toso, A. Liscidini, D. Manstretta, P. Mendez, L. Romano, and R. Castello, "An Intuitive Analysis of Phase Noise Fundamental Limits Suitable for Benchmarking LC Oscillators," *IEEE J. Solid-State Circuits*, vol. 49, no. 3, pp. 635–645, Mar. 2014.

[23] A. Mazzanti and P. Andreani, "Class-C harmonic CMOS VCOs, with a general result on phase noise," *IEEE J. Solid-State Circuits*, vol. 43, no. 12, pp. 2716–2729, Dec. 2008.

[24] M. Babaie and R. B. Staszewski, "A class-F CMOS Oscillator," *IEEE J. Solid-State Circuits*, vol. 48, no. 12, pp. 3120–3133, Dec. 2013.

[25] A. Hajimiri and T. H. Lee, "A general theory of phase noise in electrical oscillators," *IEEE J. Solid-State Circuits*, vol. 33, no. 2, pp. 179–194, Feb. 1998.

[26] Y.-H. Chillara, B. Liu, A. Wang, M. Ba, M. Vidojkovic, K. Philips, H. de Groot, and R. Staszewski, "An 860μW 2.1-to-2.7 GHz all-digital PLL-based frequency modulator with a DTC-assisted snapshot TDC for WPAN (Bluetooth Smart and Zigbee) applications," *IEEE International Solid-State Circuits Conference Digest of Technical Papers (ISSCC)*, 2014, pp. 172–173.

[27] J. Bank, "A harmonic-oscillator design methodology based on describing functions," Ph.D. dissertation, Chalmers Univ. Technology, Göteborg, Sweden, 2006.

[28] P. Andreani, X. Wang, L. Vandi, and A. Fard, "A study of phase noise in Colpitts and LC-tank CMOS oscillators," *IEEE Journal of Solid-State Circuits*, vol. 40, no. 5, pp. 1107–1118, May 2005.

[29] P. Andreani and A. Fard, "More on the phase noise performance of CMOS differential pair LC-tank oscillators," *IEEE Journal of Solid-State Circuits*, vol. 41, no. 12, pp. 2703–2712, Dec. 2006.

[30] A. Liscidini, L. Fanori, P. Andreani, and R. Castello, "A 36 mW/9 mW power-scalable DCO in 55 nm CMOS for GSM/WCDMA frequency synthesizers," *IEEE International Solid-State Circuits Conference Digest of Technical Papers (ISSCC)*, 2012, pp. 348–349.

[31] M. Garampazzi, P. M. Mendes, N. Codega, D. Manstretta, and R. Castello, "Analysis and Design of a 195.6 dBc/Hz Peak FoM P-N Class-B Oscillator With Transformer-Based Tail Filtering," *IEEE Journal of Solid-State Circuits*, vol. 50, no. 7, pp. 1657–1668, Jul. 2015.

[32] A. Mazzanti and P. Andreani, "A push-pull class-C CMOS VCO," *IEEE Journal of Solid-State Circuits*, vol. 48, no. 3, pp. 724–732, Mar. 2013.

[33] L. Fanori and P. Andreani, "A high-swing complementary class-C VCO," *Proceedings of European Solid-state Circuits Conference (ESSCIRC)*, 2013, pp. 407–410.

[34] L. Fanori and P. Andreani, "Class-D CMOS oscillators," *IEEE Journal of Solid-State Circuits*, vol. 48, no. 12, pp. 3105–3119, Dec. 2013.

[35] A. Mazzanti and A. Bevilacqua, "On the phase noise performance of transformer-based CMOS differential-pair harmonic oscillators," *IEEE Transactions on Circuits and Systems I, Reg. Papers*, vol. 62, no. 9, pp. 2334–2341, Sept. 2015.

[36] D. Murphy, J. J. Rael, A. A. Abidi "Phase noise in LC oscillators: A phasor-based analysis of a general result and of loaded Q" *IEEE Transactions on Circuits and Systems I, Reg. Papers*, vol. 57, no. 6, pp. 1187–1203, Jun. 2010.

[37] E. Klumperink, S. Gierkink, A. van der Wel, and B. Nauta, "Reducing MOSFET 1/f noise and power consumption by switched biasing," *IEEE Journal of Solid-State Circuits*, vol. 35, no. 7, pp. 994–1001, Jul. 2000.

[38] M. Shahmohammadi, M. Babaie, and R. B. Staszewski, "A 1/f noise upconversion reduction technique applied to class-D and class-F oscillators," *IEEE International Solid-State Circuits Conference Digest of Technical Papers (ISSCC)*, 2015, pp. 444–445.

[39] M. Shahmohammadi, M. Babaie, and R. B. Staszewski, "A 1/f Noise upconversion reduction technique for voltage-biased RF CMOS oscillators," *IEEE Journal of Solid-State Circuits*, vol. 51, no. 11, pp. 2610–2624, Nov. 2016.

[40] K. Kwok and H. C. Luong, "Ultra-low-voltage high-performance CMOS VCOs using transformer feedback," *IEEE Journal of Solid-State Circuits*, vol. 40, no. 3, pp. 652–660, Mar. 2005.

[41] Y. Yoshihara, H. Majima, and R. Fujimoto, "A 0.171-mW, 2.4-GHz Class-D VCO with dynamic supply voltage control," *Proceedings of European Solid-state Circuits Conference (ESSCIRC)*, 2014, pp. 339–342.

[42] N. Pourmousavian, F.-W. Kuo, T. Siriburanon, M. Babaie, and R. B. Staszewski, "A 0.5-V 1.6-mW 2.4-GHz fractional-N all-digital PLL for Bluetooth LE with PVT-Insensitive TDC using switched-capacitor doubler in 28-nm CMOS," *IEEE Journal of Solid-State Circuits (JSSC)*, vol. 53, no. 9, pp. 2572–2583, Sept. 2018.

[43] F.-W. Kuo, S. Binsfeld Ferreira, H.-N. R. Chen, L.-C. Cho, C.-P. Jou, F.-L. Hsueh, I. Madadi, M. Tohidian, M. Shahmohammadi, M. Babaie, and R. B. Staszewski, "A Bluetooth low-energy transceiver with 3.7-mW all–digital transmitter, 2.75-mW high-IF discrete-time receiver, and TX/RX switchable on-chip matching network," *IEEE Journal of Solid- State Circuits (JSSC)*, vol. 52, no. 4, pp. 1144–1162, Apr. 2017.

7

Tuning Range Extension of an Oscillator Through CM Resonance

In this chapter, we introduce a method to broaden a tuning range of a CMOS LC-tank oscillator without sacrificing its area. The extra tuning range is achieved by forcing a strongly coupled transformer-based tank into a common-mode resonance at a much higher frequency than in its main differential-mode oscillation. The oscillator employs separate active circuits to excite each mode but it shares the same tank, which largely dominates the core area but is on par with similar single-core designs. The tank is forced in common-mode oscillation by two injection locked Colpitts oscillators at the transformer's primary winding, while a two-port structure provides differential-mode oscillation. An analysis is also presented to compare the phase noise performance of the dual core oscillator in common-mode and differential-mode excitations. A prototype implemented in digital 40 nm CMOS verifies the dual mode oscillation and occupies only 0.12 mm^2 and measures 56% tuning range.

7.1 Introduction

Oscillator design for multi-mode multi-band (e.g., Fourth or Fifth Generation (4G/5G) cellular) applications demands wide tuning range (TR) while ensuring sufficiently low phase noise (PN) for a range of targeted frequency bands. The maximum achievable TR of a traditional single-core LC-tank oscillator is limited at 35%–40% by a C_{on}/C_{off} capacitance tuning ratio of its switched-capacitor network, further constrained by large size of its switches needed to prevent deterioration of the LC tank's quality (Q)-factor. For example, the Q-factor of a switched-capacitor network in a 40 nm technology is about

80 at 4 GHz resonant frequency when $C_{on}/C_{off} = 2$. For an inductor's Q-factor of 15 at this frequency, the tank's equivalent Q-factor reduces to 12.6.

The most straightforward solution seems to be designing two *separate* oscillators [1,2] at the expense of large area, and the need for high-frequency source-selecting multiplexers, which increase power consumption and noise floor. A system-level local oscillator (LO) solution in [3] uses a single 40-GHz oscillator followed by a ÷2 divider and an LC-tank mixer to generate 20 and 30 GHz LO signals. However, the extra mixer costs significant power and area as well as it produces spurs. Another attempt is to decrease the area of a *two-core* oscillator by placing one inductor underneath the other [4, 5]. However, the top inductor has to be very large so that the other one can be placed at its center without degrading the top inductor's quality factor. Therefore, the oscillator area is still considerably larger than that of a single-tank oscillator.

Employing switched resonator tanks, in which the tank's inductance is controlled by turning on/off interconnecting switches, is another TR expanding technique [6–13]. However, the switches' resistance limits the tank's Q-factor, thus degrading the oscillator PN [14]. Transformer-based dual-band oscillators [15, 16] offer wide but not continuous tuning range. A switched-shielded transformer [17] is another method to increase the oscillator's tuning range but it appears effective only at mm-wave frequencies. A shielded inductor [18] with a shorting switch is inserted between two windings of a transformer [17]. The coupling factor between the windings changes as to whether the current is flowing in the shielded inductor or not. This transformer is not large; however, its inductors' quality factor gets compromised. Consequently, this range-increasing technique is interesting for mm-wave applications where the tank's quality factor is rather limited by the capacitive part; however, for the single-GHz RF frequencies, the degradation of the tank's Q-factor would seem to be excessive.

Recent works on mode-switching oscillators significantly improve the PN versus TR trade-off [19–21]; however, they *do not* improve the TR versus die area trade-off. For example, Li et al. [20] switches between resonant modes (even/odd) of two capacitively and magnetically coupled LC resonators, as shown in Figure 7.1(a). Strong magnetic coupling enhances the difference between the two resonant frequencies; hence, a continuous TR extension calls for a low coupling factor, such that the transformer ends up to be quite large. Unfortunately, the recent CMOS technology nodes (28 nm and, to a lesser extent, 40 nm) have brought about very tough minimum metal-density

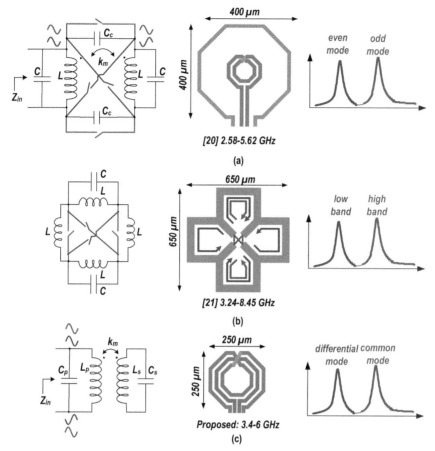

Figure 7.1 LC tanks for wide tuning range: (a) resonant mode switching technique [20]; (b) band switching technique [21]; (c) introduced technique in [33, 34].

requirements; therefore, the inductors and transformers should be filled with a lot of dummy metal pieces [22]. This has negative consequences on inductors as resistive losses due to eddy currents in the dummy fills degrade the Q-factor. And, that is in addition to increasing the parasitic capacitance, thus narrowing the TR. The losses are even more severe in the weakly coupled transformers. The spacing between their primary and secondary windings is larger (see Figure 7.1(a)) and must be filled with dummy metal pieces, but it is precisely where the magnetic flux is concentrated the most.

In [21], as shown in Figure 7.1(b), four identical inductors are coupled through four mode-switching transistors, providing two oscillation bands. In a *low-band* oscillation mode, there is no AC current flow possibility in two of these inductors (see Figure 7.1(b)); however, in a *high-band* mode, the AC current can flow in all the inductors. Thus, the effective inductance value in each band could be controlled. Obviously, the four inductors significantly increase the area.

Considering that not all applications require as stringent PN performance as does cellular wireless, we concentrate in this chapter on maintaining the die area similar to that of a single LC-tank oscillator, while significantly improving the TR and keeping a reasonable PN performance. The single-tank oscillator employs a strongly coupled transformer-based tank and forces the tank to oscillate either in a differential mode (DM) or common mode (CM); see Figure 7.1(c) [33, 34]. The DM oscillation provides the TR equivalent of a single-tank oscillator. The TR is then *extended* by the CM oscillation. The oscillator has two separate active circuits to excite each mode. However, since the passive part is shared in both modes, the die area is comparable to that of a typical narrow TR oscillator.

In Section 7.2, we briefly analyze the mode-switching oscillator introduced in [20]. Section 7.3 describes how the transformer-based tank can exhibit both DM and CM resonances. Section 7.4 describes a circuit implementation of the single-tank two-core oscillator that excites one of these resonances at a time. Section 7.5 shows measurement results.

7.2 Mode-Switching Oscillator

As we mentioned before, in this technique, two capacitively and magnetically coupled LC resonators are replaced a simple resonator to widen oscillator bandwidth. The input impedance of the transformer-based tank, shown in Figure 7.2(a), has a fourth-order polynomial denominator and shows two resonant frequencies,

$$\omega_{L,H}^2 = \frac{1 + X \pm \sqrt{(1 + X)^2 - 4X\left(1 - k_m^2\right)}}{2\left(1 - k_m^2\right)}\omega_2^2, \qquad (7.1)$$

where $\omega_1^2 = \frac{1}{L_1 C_1}$, $\omega_2^2 = \frac{1}{L_2 C_2}$ and $X = \frac{L_2 C_2}{L_1 C_1}$. The oscillator built around a transformer tank can excite ω_L or ω_H at a time to expand its tuning range. However, the different impedances of these resonances (see Figure 7.2(b)) results in a large gap in PN performance of the oscillator in two modes.

A tank can also be capacitively coupled as shown in Figure 7.2(c). Two sides of the transformer can be forced to oscillate either in phase or 180° out of phase with the help of four switches (see Figure 7.3(a)) [20]. When

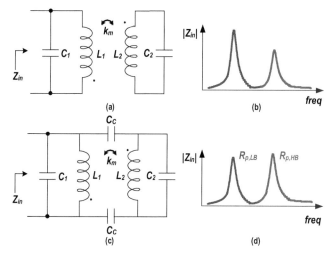

Figure 7.2 (a) Transformer-based tank and (b) its input impedance; (c) capacitively coupled transformer-based tank and (d) its input impedance.

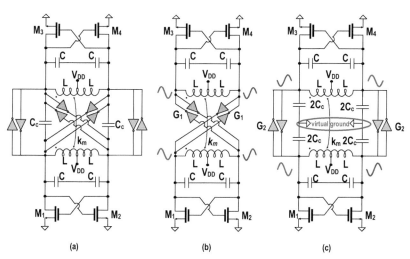

Figure 7.3 (a) Simplified schematic of DCO; (b) DCO operates in HB; and (c) LB.

oscillation is in phase, high-frequency band (HB), the coupling capacitor cannot be seen. Assuming $L_1 = L_2 = L$ and $C_1 = C_2 = C$,

$$\omega_{HB} = \frac{1}{\sqrt{(1 - k_m)LC}}. \tag{7.2}$$

However, if the two sides of the transformer are forced to oscillate out of phase, there will be a virtual ac ground in the middle of C_C, as shown in Figure 7.3(c). Therefore, the oscillator is switched to the low-frequency band (LB). The output frequency is obtained as follows:

$$\omega_{LB} = \frac{1}{\sqrt{(1 + k_m)L(C + C_C)}},$$

(7.3)

where C_C is the coupling capacitance between the two windings. A low coupling factor of the transformer, k_m, ensures that the separation between high-band and low-band frequencies is in a way that a continuous oscillation is possible. The equivalent parallel resistance of the two modes of the resonators can be found as follows [20]:

$$R_{p,HB} \approx \frac{(1 - k_m) L}{C \cdot r_s}$$

(7.4)

$$R_{p,LB} \approx \frac{(1 + k_m) L}{(C + C_c) \cdot r_s},$$

(7.5)

in which r_s is the equivalent series resistance of the primary and secondary inductances. These four design parameters, k_m, C_C, C, and L, are used to design an oscillator with continuous tuning range and some frequency overlap between the oscillation modes, while making possible $R_{p,HB} \approx R_{p,LB}$ (Figure 7.2(d)) to ensure balanced performance in the two modes.

The coupled tank resonates at one of these modes (bands) depending on the G_1 and G_2 transconductances states (see Figure 7.3(a)). When G_1's are on and G_2's are off, two sides of the tank oscillate at the same phase. In the opposite state, G_1's are off and G_2's are on, so the two sides of the resonator oscillate out of phase. In order to avoid the frequency discrepancy between HB and LB, C_C and k_m are chosen to provide some frequency overlap between the two oscillation bands and also assure almost equable phase noise performance in both modes [20]. The transconductances are designed as differential cells as is shown in Figure 7.4.

We designed a wide tuning range oscillator with this technique. It employs a transformer with $L = 700$ pH and $k_m = 0.18$ [32]. The transformer characteristics are shown in Figure 7.5(a–c). L_1 and L_2 are well designed to have more or less the same inductance. Although L_2 is considerably larger than L_1, however its quality, Q_2, factor is still 1.7 times less than Q_1.

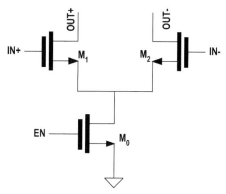

Figure 7.4 Differential transconductance schematic.

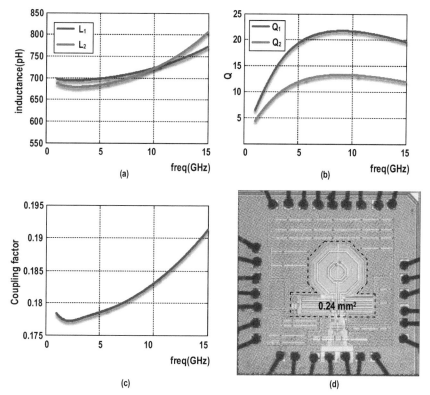

Figure 7.5 (a) Inductance and (b) quality factor of the transformer's primary and secondary winding. (c) The coupling factor. (d) Chip micrograph.

The oscillator is designed and realized in SMIC 40 nm 1P7M CMOS process. V_{DD} is chosen to be 0.6 V and the oscillation frequency is 3.6–5.02 GHZ (32% tuning range) in LB and 4.6–6.94 GHz (40% tuning range) in HB, resulting in a total of 65% tuning range. The PN performance of the f_{max}, f_{mid}, and f_{min} in LB and HB modes are shown in Figure 7.6 and Figure 7.7, respectively.

The M_1–M_4 transistor sources are connected to ground; consequently, the amount of tank's current harmonic is relatively large. In agreement with our discussion in Chapter 5, the $1/f^3$ corner is relatively large in this oscillator. For the same reason, the frequency pushing of this oscillator is also relatively high as is measured and shown in Figure 7.8.

The chip micrograph is shown in Figure 7.5(d). Active die area is about 0.24 mm^2 which is about two times larger than the rest of single tank oscillators we studied so far in this book. In Section 7.3, we study in detail a dual mode wide tuning range oscillator with an area of a single tank oscillator.

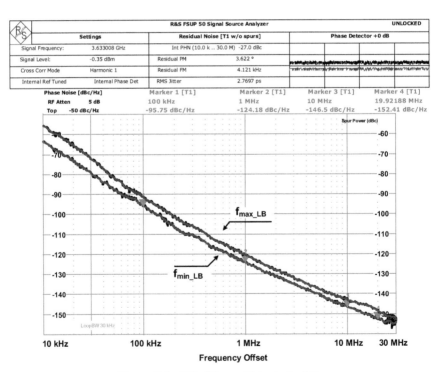

Figure 7.6   PN of the oscillator in the LB.

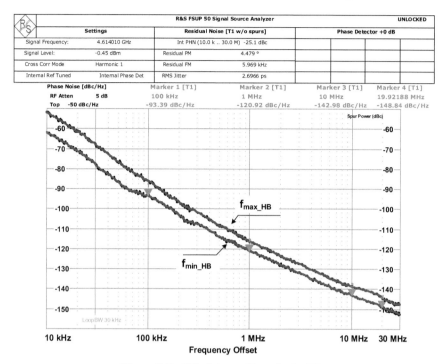

Figure 7.7 PN of the oscillator in the HB.

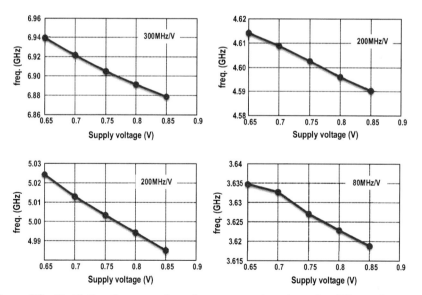

Figure 7.8 Oscillation frequency dependency on supply voltage for different frequency bands.

7.3 Common-Mode Resonances

A transformer-based tank, depicted in Figure 7.9(a), exhibits two DM resonant frequencies. If this transformer possesses a strong magnetic coupling factor, k_m, its leakage inductance would be small and so the second DM resonant frequency would be much higher than the main one. Consequently, we would not get a continuous extension of the TR by forcing the oscillation at the second DM resonant frequency. On the other hand, in order for the transformer size to be not much larger than that of an inductor,

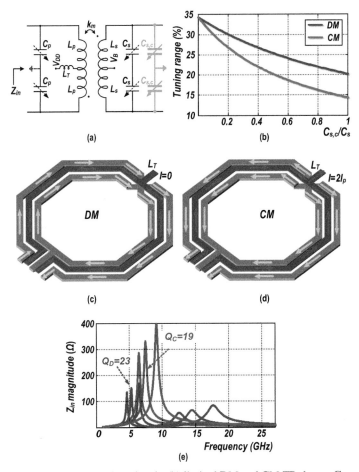

(a)

(b)

(c)

(d)

(e)

Figure 7.9   (a) A transformer-based tank; (b) limited DM and CM TR due to C_s, c; 1:2 turn transformer: (c) DM excitation; (d) CM excitation; and (e) tank's input impedance.

$k_m > 0.6$ appears a necessary condition. With this constraint, the first resonance can be estimated as [24]

$$\omega_{0,DM} \approx \frac{1}{\sqrt{L_pC_p + L_sC_s}}, \tag{7.6}$$

where L_p and C_p are primary, and L_s and C_s are secondary windings' inductances and capacitances. The approximation error of (7.6) from the exact resonant frequency (Equation (5) in [24]) is less than $+6\%$ for $k_m \geq 0.7$.

Abandoning the hope of exploiting the second DM resonance, suppose now this tank is excited by CM signals, and, for now, we assume that primary and secondary winding inductances and k_m are similar in DM and CM excitations. CM signals cannot see the differential capacitors; thus, the tank can only exhibit CM resonances when these capacitors are single-ended. If this tank were to employ only single-ended primary and *differential* secondary capacitors, the secondary winding inductances and capacitances would not affect the CM characteristics of the tank, e.g., resonant frequency. This tank will show a single CM resonance at

$$\omega_{0,CM} \approx \frac{1}{\sqrt{L_pC_p}}. \tag{7.7}$$

The difference between the CM and DM resonance frequencies, i.e., Equations (7.7) and (7.6), suggests a new possibility for extending the tuning range toward higher frequencies, provided we can build an oscillator around this transformer-based tank that can excite it with either DM or CM signals, without adding any bulky passive components. To investigate how much tuning range we can expect from a single tank, we assume that the tank employs a switched capacitor bank with a 2:1 capacitance switching ratio:

$$C_{p,max}/C_{p,min} = C_{s,max}/C_{s,min} = 2. \tag{7.8}$$

This ratio should guarantee a sufficiently high Q-factor of switched-capacitors in recent CMOS technologies. With this assumption, $f_{max}/f_{min} = \sqrt{2}$ in both modes and, thus, both DM and CM resonant frequencies (Equations (7.6) and (7.7)) will tune by $2(\sqrt{2} - 1)/(\sqrt{2} + 1) = 34.3\%$. To avoid any gaps between the DM and CM tuning ranges, at least $\omega_{CM,low} = \omega_{DM,high}$. Hence,

$$L_pC_{p,max} = L_sC_{s,max}. \tag{7.9}$$

With these conditions, the resonant frequency could theoretically cover an octave while going from DM to CM oscillations. Practically, C_{max}/C_{min} has to be >2 due to parasitics and difficulty with controlling the precise overlap between the DM and CM resonances.

One limiting factor in the tuning range of such an oscillator is the single-ended parasitic capacitance throughout the secondary winding side. If the CM coupling factor, $k_{m,c}$, were hypothetically similar to the DM one, $k_{m,d}$, and $k_{m,c} = k_{m,d} > 0.6$, then the CM resonance would shift down to $\omega_{0,CM} = 1/\sqrt{L_pC_p + L_sC_{s,c}}$, where $C_{s,c}$ is the total of single-ended capacitances on the secondary side (Figure 7.9(a)). At the same time, the DM resonance would also shift down to $\omega_{DM} = 1/\sqrt{L_pC_p + L_sC_s + L_sC_{s,c}}$. Interestingly, satisfying the overlap between CM and DM oscillations with the condition in (7.8) results in the same constraint as (7.9). However, the fixed parasitic capacitance, $C_{s,c}$, degrades the CM oscillation tuning range more than it degrades the DM oscillation tuning range; see Figure 7.9(b).

A 1:2 turns-ratio transformer, which has distinctly different characteristics in DM and CM excitations, relieves such a degradation. Figures 7.9(c,d) show this transformer when its primary is excited, respectively, with DM or CM signals. In the DM excitation, the induced currents at the two sides of the secondary winding circulate constructively in the same direction, thus creating a strong coupling factor between the transformer windings, while in the CM excitation these induced currents cancel each other within each full turn of the secondary winding (i.e., from the secondary's terminal to the secondary's center-tap), leading to a weak coupling factor [23]. This weak $k_{m,c}$ can be interpreted as the secondary winding not being seen from the primary and, therefore, the secondary's single-ended capacitors have an insignificant effect on the tank's CM resonant frequency.

Assuming the capacitor bank is almost ideal, at least compared to the lossy inductors represented by the r_p and r_s equivalent series resistances of the primary/secondary windings, CM resonance has the quality factor of $Q_{CM} = Q_p = L_p\omega/r_p$, which is similar to that of an inductor-based tank. The high Q-factor of this resonance indicates that with an appropriate active circuitry, the CM oscillation of a reasonable quality would be possible. The DM and CM input impedances of this tank are shown in Figure 7.9(e). The single-ended switched capacitors require two switches to provide a ground connection in the middle, which results in a 50% lower Q-factor as compared to a differential switched capacitor with the same switch size. This would appear as a disadvantage of our new technique; however, that is not the case. Let us compare the tuning range of a typical inductor-based tank oscillator employing the differential capacitor bank with our transformer-based tank oscillator employing the single-ended primary and the differential

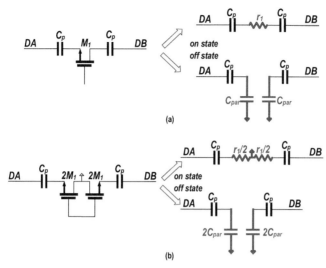

Figure 7.10 (a) Differential and (b) single-ended capacitor banks.

secondary capacitor banks. The equivalent capacitance of this bank varies from $C_{on,D} = C_p$ to $C_{off,D} = \frac{C_p C_{par}}{C_p + C_{par}}$, where C_{par} is the parasitic capacitance of the switch (see Figure 7.10(a)). For a typical $C_{off,D}/C_{on,D}$ value of 0.5 ($C_{par} = C_p$), the inductor-based oscillator employing this tank would exhibit $f_{max}/f_{min} = \sqrt{2}$.

The width of each switch in the single-ended switched-capacitor bank should be twice the width of each differential counterpart for the same Q-factor. Consequently, $C_{off,C} = \frac{2C_p C_{par}}{C_p + 2C_{par}} = \frac{2}{3}C_p$ (see Figure 7.10(b)). Employing this capacitor bank in a transformer-based tank at the primary winding and employing the differential bank at the secondary winding, and benefiting from the impedance transformation of the 1:2 turns-ratio transformer ($L_s/L_p \approx 3$), results in $\frac{f_{max}}{f_{min}} = \sqrt{1.9}$, which is very close to the inductor-based tank tuning range.

7.4 Novel Wide Tuning Range Oscillator

7.4.1 Dual-Core Oscillator

Forcing the transformer-based tank to resonate in DM is quite straightforward. The oscillator can be realized as a one-port or a two-port structure [25, 26]. However, only the two-port structure will guarantee a reliable

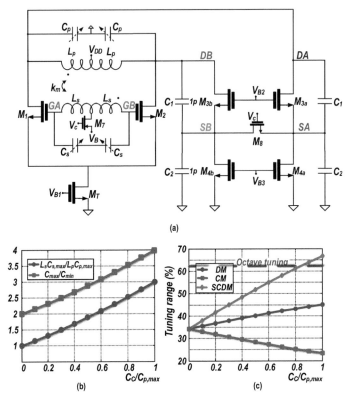

Figure 7.11   Dual core oscillator: (a) schematic; (b) overlap and octave oscillation conditions; and (c) tuning range.

start-up at the first DM resonance [24], thus preventing the mixed DM oscillation. A separate active circuit is now needed to force the tank into the CM resonance. Colpitts and Hartely topologies are two well-known examples of single-ended oscillators. Invoking our ground principle of sharing the *same* tank by the active CM and DM circuits, the Coplitts structure is consequently chosen. To improve the PN, two mutually injection-locked Colpitts oscillators share the primary inductor. The schematic of the novel dual core oscillator is shown in Figure 7.11(a). To avoid the dual oscillation, only one active circuit core is turned on at a time.

The left side of Figure 7.11(a) is the two-port DM oscillator. In this mode, $V_{B2} = V_{B3} = 0V$, M_7 switch is on biasing $M_{1,2}$, while M_8 switch is off. The waveforms are shown in Figure 7.12(a,b). The transformer has the 1:2 turns ratio and its gain reduces the $M_{1,2}$ noise upconversion to PN, and also results in a larger gate voltage compared to drain voltages, which facilitates oscillation start up.

The right-hand side of the oscillator schematic are two locked single-ended Colpitts oscillators. M_8 switch is now turned on to ensure the in-phase operation of the two Colpitts oscillators, without which the two cores might exhibit an arbitrary phase shift. In this mode, $V_{B1} = 0V$ to turn off the differential oscillation. M_7 switch is also off to minimize the CM inductive loading on the primary winding by the secondary one. Both single-ended oscillators start at the same frequency but could be slightly out of phase; subsequently, they lock to each other and there is no phase shift between them. The locking of the two oscillators gives an additional 3-dB PN improvement. Waveforms are shown in Figure 7.12(c,d).

Note that an attempt of simplifying the CM structure by removing M_8 and permanently shorting the sources of M_3 transistors would be detrimental to the DM tuning range. While obviously the DM oscillation would still work – M_3 transistors are off in this mode – the extra capacitance C_{fix} due to the CM circuitry seen by C_p would be larger. With M_8 off, DA/DB node sees

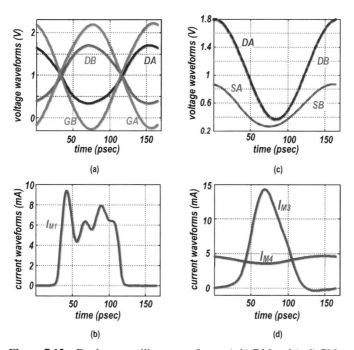

Figure 7.12 Dual core oscillator waveforms: (a,b) DM and (c,d) CM.

$C_{fix} = C_1 C_2/(C_1 + C_2)$, but when M_3 sources are shorted, that capacitance raises to $C_{fix} = C_1 > C_1 C_2/(C_1 + C_2)$. Furthermore, an attempt of moving M_8 from the SA/SB source nodes of M_3 to the DA/DB drain nodes would likewise increase the effective parasitic capacitance of M_8.

The C_1 and C_2 capacitors are necessary to create a negative resistance for the Colpitts oscillators; however, they are limiting the tuning range in both modes. In their presence, (7.8) and (7.9) are not valid anymore for the overlap and octave tuning. Assuming the same capacitance variation range on the primary and secondary sides, $C_{p,max}/C_{p,min} = C_{s,max}/C_{s,min}$, the octave tuning requirement is now

$$\frac{L_s C_{s,max}}{L_p C_{p,max}} = 3\frac{C_C}{C_{p,max}} + 4\frac{C_{p,min}}{C_{p,max}} - 1, \tag{7.10}$$

where $C_C = C_1 C_2/(C_1 + C_2)$. The minimum overlap condition, $f_{DM,max} = f_{CM,min}$, dictates

$$\frac{C_{p,max}}{C_{p,min}} = 1 + \frac{L_s C_{s,max}}{L_p C_{p,max}}. \tag{7.11}$$

Figure 7.11(b) shows how the required C_{max}/C_{min} increases with C_C/C_p ratio. Satisfying (7.11) and (7.10) in the presence of C_C also unbalances the DM and CM tuning range, as shown in Figure 7.11(c). For a certain value of C_C, the required $C_{p,max}/C_{p,min}$ ratio can become prohibitively large, likely leading to the Q-factor degradation. In practice, $C_{s,max}/C_{s,min}$ and $C_{p,max}/C_{p,min}$ should not be necessarily equal. The secondary-winding capacitor ratio in this design is chosen to be larger than at the primary side due to the tougher Coplitts oscillator start-up conditions.

7.4.2 Phase Noise Analysis

Ideally, a wide TR oscillator would have a comparable PN performance in both oscillation modes. In this section, we investigate the PN of the dual core oscillator and then compare the two modes.

The linear time-variant model [28] suggests

$$\mathcal{L}(\Delta\omega) = 10\log_{10}\left(\frac{kT}{R_t N q_{max}^2 (\Delta\omega)^2} \cdot F\right), \tag{7.12}$$

where k is Boltzmann's constant, T is temperature, R_t is the equivalent parallel resistance of the tank, and q_{max} is the maximum charge displacement across the equivalent capacitance in parallel to R_t. N is the number of resonators, which is 2 here in both DM and CM oscillators. F, the oscillator's effective noise factor, is

$$F = \sum_i \frac{N \cdot R_t}{2kT} \cdot \frac{1}{2\pi} \int_0^{2\pi} \Gamma_i^2 (\phi) \, \overline{i_{n,i}^2(\phi)} \, d\phi, \qquad (7.13)$$

in which Γ_i is the ISF of the ith noise source. The relevant ISF of noise sources associated with a sinusoidal waveform oscillator can be estimated by a $\pi/2$ phase shifted sinusoidal function, $\Gamma = \frac{sin(\phi)}{N}$, where $\phi = \omega_0 t$ [27]. Here, we try to find the noise factors of different noise sources in the dual core oscillator.

The noise sources of the Colpitts oscillator are R_t, M_3, and M_4. R_t in the CM oscillation is the parallel resistance of the primary winding, R_p. It is insightful to refer every noise source and nonlinearity back to the tank, as it is demonstrated step-by-step in Figure 7.13. The negative conductance between DA and SA nodes is

$$g_n = \frac{i_{d3}}{v_{DA} - v_{SA}} = \frac{-g_{m3}v_{SA}}{v_{DA} - v_{SA}} = -g_{m3}\frac{C_1}{C_2}, \qquad (7.14)$$

where i_{d3} is the small-signal drain current of M_3. The equivalent negative conductance in parallel with the tank is found as

$$G_n = \left(\frac{C_2}{C_1 + C_2}\right)^2 \cdot g_n = -g_{m3}\frac{C_1 C_2}{(C_1 + C_2)^2}. \qquad (7.15)$$

With a similar derivation, M_3 channel resistance is referred to the tank as

$$R_{ds3} = r_{ds3} \left(\frac{C_1 + C_2}{C_2}\right)^2. \qquad (7.16)$$

To sustain the oscillation, the average dissipated power in the tank loss and R_{ds3} should be equal to the average power delivered by the negative resistance, which leads to the condition:

$$G_{mEF3} = \frac{1}{n(1 - n)} \cdot \frac{1}{R_p} + \frac{1 - n}{n} \cdot G_{dsEF3}, \qquad (7.17)$$

where $n = C_1/(C_1 + C_2)$, $G_{mEF} = G_m[0] - G_m[2]$, and $G_{dsEF} = G_{ds}[0] - G_{ds}[2]$, in which $G_m[k]$ and $G_{ds}[k]$ are the kth Fourier coefficients of $g_m(t)$ and $g_{ds}(t)$, respectively [29]. The required G_{mEF3} is minimized for $n = 0.5$, which is chosen in this design to facilitate start-up.

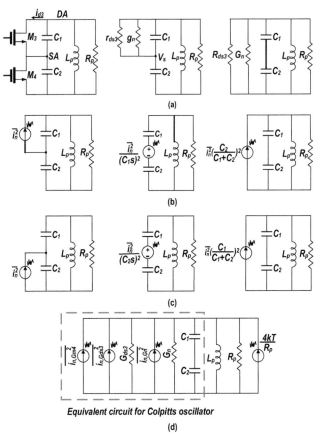

Figure 7.13   Procedures of referring the noise back to the tank from: (a) r_{ds3} and negative conductance; (b) g_m of M_3; and (c) g_m of M_4. (d) The equivalent circuit of the Colpitts oscillator.

To refer the current noise sources to the tank, they are first converted to their Thevenin voltage source equivalents and then converted back to Norton current source equivalents, as demonstrated in Figure 7.13(b) and (c). The equivalent noise of M_3 and M_4's transconductance then becomes

$$\overline{i_{n3}^2} = 4kT\gamma g_{m3} \left(\frac{C_2}{C_1 + C_2}\right)^2, \tag{7.18}$$

$$\overline{i_{n4}^2} = 4kT\gamma g_{m4} \left(\frac{C_1}{C_1 + C_2}\right)^2, \tag{7.19}$$

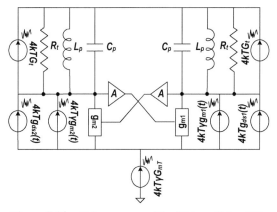

Figure 7.14 Noise sources of the DM oscillator [24].

where γ is the transistor excess noise coefficient. Assuming a sinusoidal oscillation, the tank noise factor is found as

$$F_t = \frac{2N}{2kTR_p} \cdot \frac{1}{2\pi} \int_0^{2\pi} \frac{4kT}{R_p} \frac{\sin^2(\phi)}{N^2} d\phi = 1. \qquad (7.20)$$

M_3 and M_4 noise factors are found as

$$F_{gm3} = \frac{2NR_p}{4kT\pi} \int_0^{2\pi} \frac{\sin^2(\phi)}{N^2} 4kT\gamma g_{m3}(\phi) \cdot \left(\frac{C_2}{C_1 + C_2}\right)^2 d\phi$$
$$= (1-n)^2 \gamma G_{mEF3} R_P \qquad (7.21)$$

$$F_{gds3} = \frac{2NR_p}{4kT\pi} \int_0^{2\pi} \frac{\sin^2(\phi)}{N^2} 4kT\gamma g_{ds3}(\phi) \cdot \left(\frac{C_2}{C_1 + C_2}\right)^2 d\phi$$
$$= (1-n)^2 G_{dsEF3} R_p \qquad (7.22)$$

$$F_{gm4} = \frac{2NR_p}{4kT\pi} \int_0^{2\pi} \frac{\sin^2(\phi)}{N^2} 4kT\gamma g_{m4} \cdot \left(\frac{C_1}{C_1 + C_2}\right)^2 d\phi$$
$$= n^2 \gamma G_{mEF4} R_P \qquad (7.23)$$

g_{ds4} noise is very small due to M_4 operating in a saturation region and, consequently, is disregarded in our calculations. Since g_{m4} is fairly constant throughout the period, $G_{mEF4} = g_{m4}$. To estimate the contribution of M_4 to PN, we can calculate g_{m4} as

$$g_{m4} = \frac{2I_0}{V_{gs4} - V_{th}} \approx \frac{2I_0}{V_{ds,min}}, \qquad (7.24)$$

where V_{th} is the transistor's threshold voltage. Let us assign $V_{DD}/2$ to the SA (SB) node, and $V_{DA} \approx 2I_0 R_p(1-n)$ [27],

$$g_{m4} \approx \frac{4I_0}{V_{DD} - 4n(1-n)I_0 R_p}. \tag{7.25}$$

Disregarding g_{ds4} noise contribution,

$$F_{M4} \approx F_{gm4} = \frac{4n^2\gamma R_p I_0}{V_{DD} - 4n(1-n)I_0 R_p} \approx \gamma. \tag{7.26}$$

By substituting (7.17) in (7.21), with G_{mEF3} and G_{dsEF3} numerically obtained from simulations, the total oscillator effective noise factor then will be

$$F_{CM} = R_P \left[(1-n)^2 G_{mEF3} \left(\gamma + \frac{n}{1-n} \right) + \frac{4n^2\gamma I_0}{V_{DD} - 4n(1-n)I_0 R_p} \right]$$
$$- 1 \approx 2.2\gamma + 0.2. \tag{7.27}$$

The circuit-to-phase-noise conversion of the CM oscillator is shown in Figure 7.15(a–d).

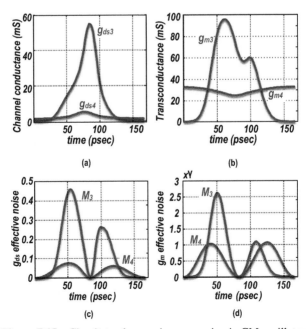

Figure 7.15   Circuit-to-phase-noise conversion in CM oscillator.

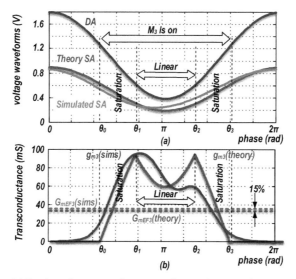

Figure 7.16 (a) Drain and source voltage waveforms. (b) g_{m3}: theory and simulations.

The noise contribution of M_3 transistor can be numerically calculated based on design parameters. For M_3, $V_d(\phi) = V_{DD} + A_C \cos(\phi)$, $V_s(\phi) \approx V_{DD}/2 + nA_C \cos(\phi)$ and $V_g = V_{B2}$. Figure 7.16 shows the M_3 operating regions during one oscillating period. At θ_0, V_s gets low enough for M_3 to turn on and enter the saturation region. When the drain voltage gets lower, M_3 enters the triode region at θ_1 and remains there till $\theta_2 = 2\pi - \theta_1$. M_3 finally turns off again at $\theta_3 = 2\pi - \theta_0$. θ_0 and θ_1 can be found from boundary conditions as

$$\theta_0 = \cos^{-1}\left(\frac{V_1}{nA_C}\right) \tag{7.28}$$

and

$$\theta_1 = \cos^{-1}\left(\frac{V_2}{A_C}\right) \tag{7.29}$$

where $V_1 = V_{B2} - V_{DD}/2 - V_{th}$ and $V_2 = V_{B2} - V_{DD} - V_{th}$.

Assuming square law,

$$g_{m3}(\phi) = \begin{cases} K(V_1 - nA_C \cos(\phi)) & \text{saturation,} \\ K(\frac{V_{DD}}{2} + (1-n)A_C \cos(\phi)) & \text{linear,} \\ 0 & \text{cut-off,} \end{cases} \tag{7.30}$$

where $K = \mu C_{ox}\left(\frac{W}{L}\right)$ is the customary designation of MOS transistor strength. G_{mEF3} now can be determined by calculating the Fourier coefficients of $g_{m3}(\phi)$. Solving the lengthy integrations results in

$$G_{mEF3} = \frac{K}{2\pi}[2V_1(\theta_1 - \theta_0) + V_{DD}(\pi - \theta_1) + nA_c\sin(\theta_0) - A_c\sin(\theta_1)$$
$$+ V_1\sin(2\theta_0) + \left(\frac{V_{DD}}{2} - V_1\right)\sin(2\theta_1) - \frac{nA_c}{3}\sin(3\theta_0)$$
$$+ \frac{A_c}{3}\sin(3\theta_1). \tag{7.31}$$

G_{mEF3} in (7.31) can be calculated by substituting θ_0 and θ_1 from (7.28) and (7.29), together with other design parameters: $V_{DD} = 1.1$ V, $V_{B2} = 1$ V, $V_{th} \approx 0.37$ V. Figure 7.16(b) shows a very good agreement (within 15%) with the simulation results.

Major noise sources of the DM oscillator are shown in Figure 7.14. A general result of the effective noise factor, assuming that the M_T thermal noise is completely filtered out, is derived in [24] as

$$2\Gamma_{t,rms}^2 \cdot \left(1 + \frac{\gamma}{A}\right) \cdot (1 + R_t G_{dsEF1}) \approx 1.6 + 0.9\gamma. \tag{7.32}$$

However, the M_T thermal noise is not completely filtered out here. To calculate the M_T's noise contribution, the tail node ISF is obtained through simulations and plotted in Figure 7.17(e). From that

$$F_{MT} = \frac{1}{2\pi}\int_0^{2\pi} 4kT\gamma g_{mT} \cdot \Gamma_{MT}^2(t)\frac{R_t}{4kT}\,dt \approx 0.5\gamma. \tag{7.33}$$

Hence, the DM oscillator noise factor is

$$F_{DM} = 2\Gamma_{t,rms}^2 \cdot \left(1 + \frac{\gamma}{A}\right) \cdot (1 + R_t G_{dsEF1})$$
$$+ \Gamma_{MT,rms}R_p G_{MTEF} \approx 1.6 + 1.4\gamma. \tag{7.34}$$

The DM oscillator circuit-to-phase-noise conversion is shown in Figure 7.17(a–d).

Substituting (7.27) and (7.34) in (7.12) at the *overlap* frequency results in

$$\mathcal{L}_{DM} - \mathcal{L}_{CM} = 10\log_{10}\left(\frac{R_t Q_p A_C^2}{R_p Q_t A_D^2} \cdot \frac{F_{DM}}{F_{CM}}\right) \approx -2.5 \text{ dB}. \tag{7.35}$$

Due to its single-ended structure and the CM resonance, the Colpitts oscillator would appear to be more sensitive to supply noise. However, that is

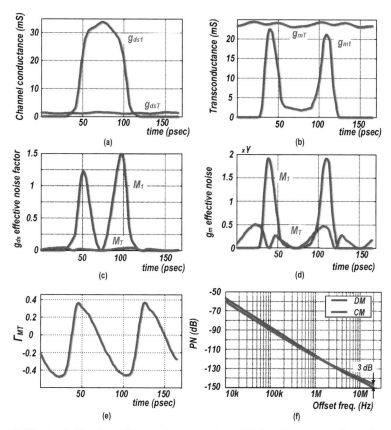

Figure 7.17   (a–d) Circuit-to-phase-noise conversion in DM oscillator; (e) tail transistor ISF; and (f) PN of CM and DM oscillators at the overlap frequency.

not the case. Supply pushing is the parameter that indicates the supply noise effect on the phase noise. Figure 7.19(e,f) demonstrates this parameter for the DM and CM oscillators, which is quite comparable, indicating that the CM oscillation does not result in higher phase noise upconversion sensitivity to the supply noise. To explain that, let us look at the actual mechanism: the oscillation frequency can be modulated by the supply noise by modulating the nonlinear voltage-dependent parasitic capacitors of the core transistors, C_{gs}. In the Colpitts oscillator, the supply voltage is connected to the core transistors' drains, which cannot modulate their C_{gs} directly. Consequently, the oscillation frequency modulation due to the supply noise is considerably reduced.

7.4.3 Center Tap Inductance

The single-ended nature of the Colpitts oscillator makes its characteristics especially sensitive to single-ended parasitics. A key parasitic that must be properly modeled and accounted for is the metal track inductance, L_T, which connects the center tap of the transformer's primary to the supply's AC-ground (see Figure 7.9(a)). At the DM excitation, the AC current will not flow into L_T; thus, the DM inductance and DM resonant frequency are independent of its value. However, at the CM excitation, the current flowing into L_T is twice the current circulating in the inductors. Consequently, the tank inductance L_p in Figure 7.9(a) is re-labeled as $L_{pd} = L_p$ in DM and $L_{pc} = L_{pd} + 2L_T$ in CM excitations. The CM oscillation frequency will be reduced to $\omega_{CM} = 1/\sqrt{(L_p + 2L_T)C_P}$. This implies that L_T must be carefully modeled and included in simulations, otherwise the increased overlap between CM and DM oscillations would severely limit the total tuning range.

Another important parasitic that is only influential in the CM oscillation is the supply loop resistance between the V_{DD} feed to the center-tap of the primary winding and the sources of M_4 transistors (see Figure 7.11), assuming sufficient decoupling capacitance on V_{DD}. This resistance is added directly to the equivalent negative resistance of the Colpitts structure and increases it from $-g_{m3}/C_1C_2\omega^2$ to $-g_{m3}/C_1C_2\omega^2 + r_b$. In our design, the average of that negative resistance at 6 GHz with $C_1 = C_2 = 1$ pF is about $-25\ \Omega$, which means the r_b parasitic resistance should be kept much smaller as to not endanger the start-up.

7.5 Experimental Results

The novel oscillator is prototyped in TSMC 40 nm 1P7M CMOS process with top ultra-thick metal. $M_{1,2}$ are (60/0.27) μm and $M_{3,4}$ are (128/0.04) μm low-V_{th} devices for safe start-up of the Colpitts oscillator. The tank employs a 1.4 nH secondary inductor with Q of 25 at 5 GHz and 0.54-nH primary inductor with Q of 17 at 5 GHz. $k_{m,DM} = 0.72$ and $k_{m,CM} = 0.29$. The transformer size is 250×250 μm^2 and the primary-to-secondary winding spacing is 5 μm. The chip micrograph and transformer characteristics are shown in Figure 7.18, respectively. The oscillator's core area is 0.12 mm^2, which is similar in size to typical narrow tuning-range oscillators. The tank is shared in the two modes of oscillation and so the output is common; hence, no further multiplexing is necessary. A comparison with other relevant wide

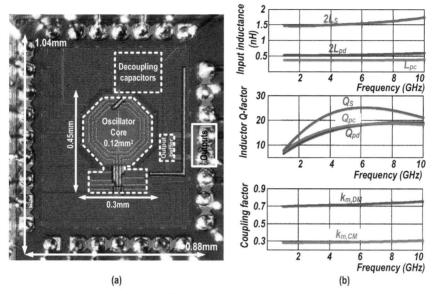

Figure 7.18 Transformer characteristics.

Table 7.1 Performance summary and comparison with relevant oscillators

		This Work		[20]		[21]		[4]		[5]		[13]		[14]
Frequency (GHz)		**3.37–5.96**		2.5–5.6		3.24–8.45		2.4–5.3		1.3–6		3.28–8.35		3.14–6.44[2]
Tuning range (%)		**55.5**		76		89		75.3		128		87.2		69[2]
V_{DD} (V)		**1**		0.6		0.8		0.4		1.5		1.6		1.2
Technology		**40 nm**		65 nm		40 nm		65 nm		130 nm		130 nm		180 nm
OSC core area		**0.12 mm²**		0.29 mm²		0.43 mm²		0.25 mm²		0.295 mm²		0.1 mm²		0.35 mm²
		f_{min}	f_{max}	f_{min}	f_{max}	f_{min}	f_{max}	f_{min}	f_{max}	f_{min}	f_{max}	f_{min}	f_{max}	f_{mid}
P_{DC} (mW)		**16**	**12.5**	14.1	9.9	16.5	14	6	4.4	4.35[1]	9.15[1]	15.4	6.5	8.8
PN	100 kHz	**–103**	**–90**	–101.1	–89	–109	–91	–98	–86	NA	NA	–96	NA	–92
(dBc/Hz)	10 MHz	**–149.7**	**–137.8**	–151.9	–145.8	–150	–142	–149	–139	–135	–132	–142	–137.2	–140
FoM[†]	100 kHz	**181.8**	**174.5**	177.6	174	187	178.1	177.8	174.1	NA	NA	174.4	NA	175.4
(dB)	10 MHz	**188.2**	**182.3**	188.4	190.8	188	189.1	188.8	187	171	178	180.4	187.5	183.4
FoMA[††]	100 kHz	**191**	**183.7**	182.9	179.4	190.7	181.7	183.8	180.1	NA	NA	184.4	NA	180
(dB)	10 MHz	**197.4**	**191.5**	193.7	196.2	191.7	192.7	194.8	193	176.2	183.2	190.4	197.5	188
FoMAT[†††]	100 kHz	**205.6**	**198.6**	200.5	197	209.7	200.7	201.3	197.6	NA	NA	203.3	NA	196.8
(dB)	10 MHz	**212.3**	**206.4**	211.3	213.8	210.7	219.7	212.3	210.5	198.3	205.4	209.3	216.3	204.8

†$FoM = |PN| + 20\,log_{10}(\omega_0/\Delta\omega) - 10\,log_{10}(P_{DC}/1mW)$.

††$FoMA = |PN| + 20\,log_{10}(\omega_0/\Delta\omega) + 10log(1mm/A) - 10\,log_{10}(P_{DC}/1mW)^2$.

†††$FoMAT = |PN| + 20\,log_{10}(\omega_0/\Delta\omega) + 20\,log_{10}(TR/10) + 10log(1mm^2/A) - 10\,log_{10}(P_{DC}/1mW)$.

[1] Including bias circuitry.

[2] Before frequency division.

tuning-range oscillators is summarized in Table 7.1. This oscillator is smaller by at least a factor of 2. The oscillators are tuned via 4-bit switched MOM capacitor banks at the primary and secondary. According to post-layout circuit-level simulations, the tuning range is 46% in DM and 20% in CM, with a 100 MHz overlap, giving the total TR of 63%.

However, measurements show that DM oscillator is tunable between 3.37 and 5.32 GHz (45% TR) and the CM oscillator is tunable between 5.02 and 5.96 GHz (17% TR) and the overlap between the DM and CM oscillations is wider than expected, resulting in a tuning range of 55.5%.

Figure 7.19 shows PN at f_{max} and f_{min} frequencies of the DM and CM oscillations. In both modes, V_{DD} is 1.1 V. Figure 7.19 also reports the PN and FoM of this oscillator over the tuning range. The FoM increases from 188.2 to 189.4 dB in the DM and from 181.3 to 182.3 dB in the CM tuning ranges. The PN in the CM mode is worse than that in the DM mode, but it is worth mentioning that not all applications demand ultra-low phase noise in all bands and channels uniformly.

Table 7.1 also compares FoMA, introduced in [31], of this oscillator with other relevant oscillators. The DM oscillator shows the best FoMA and the CM oscillator's FoMA is comparable with the other state-of-the-art oscillators.

7.5.1 Supply and Ground Routing Inductances and Losses

The measurement results deviate from the simulations and theory in two ways. The first is the wider overlap between the DM and CM oscillation frequencies. The second is the degraded PN in the CM Colpitts oscillator. To explain the performance degradation, we first take a closer look at a layout of the transformer-based tank. As revealed in Figure 7.20, the CM inductance should also include the impedance of the current return route, from the center-tap of the primary winding to the sources of M_{4a} (M_{4b}). The de-coupling capacitors together with the RLC routing network present an equivalent impedance that is inductive but its real part adds to the circuit losses. Therefore, unless the return current path happens to resonate at the same oscillation frequency (through the equivalent inductances and decoupling capacitors along it), the CM oscillation shifts down from the expected value, which is precisely what we observe in our measurements. The DM oscillation frequency is not affected; therefore, the expected TR is decreased. Furthermore, the losses in the return path are added to the losses of the primary inductor, thus degrading the quality factor of the tank. The long return path causes the losses to be comparable to the inductor's loss and this jeopardizes the CM start-up. Furthermore, this path also partially cancels the magnetic field of the inductor, thus degrading its Q-factor. The severe PN degradation compared to the simulation results gives, thus, credence to the Q degradation of the tank.

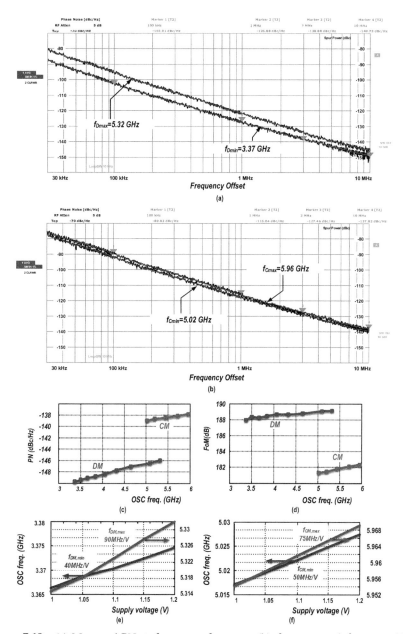

Figure 7.19 (a) Measured PN at $f_{DM,max}$, $f_{DM,min}$; (b) $f_{CM,max}$ and $f_{CM,min}$. Measured (c) PN and (d) FoM at 10-MHz offset across TR. Frequency pushing due to supply voltage variation in (e) DM and (f) CM oscillators.

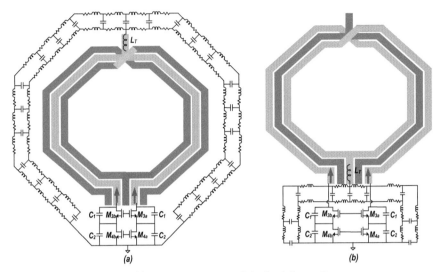

Figure 7.20   Return current path in the 1:2 transformer.

Our EM simulations predict a 0.25-Ω resistance in this path and circuit simulations show that such resistance in series with the primary inductor would degrade the CM oscillator phase noise by 4 dB. This appears to agree with our measurements.

One possible solution would be employing a 2:1 transformer. A 2-turn primary inductor will have its supply connection node very close to the transistors; therefore, the current return path would not be very long, thus minimizing the path inductance. However, in that transformer, the CM current in the two windings of the primary inductor has opposite direction, thus canceling each other's flux [30]. Consequently, the CM primary inductance would be smaller than the DM one. The spacing between the transformer windings should be chosen properly to satisfy the overlap condition for the reasonable capacitor bank $C_{\text{on}}/C_{\text{off}}$ ratios.

7.6 Conclusion

In this chapter, we have introduced a technique to extend a tuning range (TR) of an LC-tank oscillator without significantly increasing its die area. A strongly coupled 1:2 turns-ratio transformer-based tank is normally excited in a differential mode (DM), where it achieves the TR of 45% with a good FoM of 188.2–189.4 dB. The TR is extended by exciting the tank in common mode (CM) with two locked Colpitts oscillators. This oscillator is implemented in 40 nm CMOS and delivers the total TR of 55.5% while constraining the core die area to only 0.12 mm^2. Although the measured tuning range extension and phase noise (PN) in the CM mode were worse than theoretically predicted, we have identified the common cause as a current return route inductance that not only lowers the CM frequencies but also adds losses that result in a reduced Q-factor.

References

[1] M. Carusol, M. Bassi, A. Bevilacqua, and A. Neviani, "Wideband 2–16GHz local oscillator generation for short-range radar applications," *in Proc. of the IEEE European Solid-State Circuits Conference(ESSCIRC)*, 2012, pp. 353–356.

[2] J. Borremans et al., "A 86 MHz–12 GHz digital-intensive PLL for software-defined radios, using a 6 fJ/step TDC in 40 nm digital CMOS" *IEEE J. Solid-State Circuits*, vol. 45, no. 10, pp. 2116–2129, Oct. 2010.

[3] Y. Chen, Y. Pei, and D. M. W. Leenaerts, "A dual-band LO generation system using a 40GHz VCO with a phase noise of 106.8dBc/Hz at 1-MHz," *in IEEE Radio Frequency Integrated Circuits Symposium (RFIC)*, 2013, pp. 203–206.

[4] L. Fanori, T. Mattsson, and P. Andreani, "A 2.4-to-5.3 GHz dual core CMOS VCO with concentric 8-shape coils," *in IEEE Int. Solid-State Circuits Conf. Dig Tech. Papers (ISSCC)*, Feb. 2014, pp. 370–372.

[5] Z. Safarian, and H. Hashemi, "Wideband multi-mode CMOS VCO design using coupled inductors," *IEEE Trans. Circuits Syst. I, Reg. Papers*, vol. 56, no. 8, pp. 1830–1843, Aug. 2007.

[6] S.-M. Yim and K. O. Kenneth, "Demonstration of a switched resonator concept in a dual-band monolithic CMOS LC-tuned VCO," *in Proc. IEEE Custom Integr. Circuits Conf.*, 2001, pp. 205–208.

[7] N. D. Dalt, E. Thaller, P. Gregorius, and L. Gazsi, "A compact triple-band low-jitter digital LC PLL with programmable coil in 130 nm CMOS," *IEEE J. Solid-State Circuits*, vol. 40, no. 7, pp. 1482–1490, Jul. 2005.

[8] Z. Li and K. S. O, "A low-phase-noise and low-power multiband CMOS voltage-controlled oscillator," *IEEE J. Solid-State Circuits*, vol. 40, no. 6, pp. 1296–1302, Jun. 2005.

[9] D. Hauspie, E. Park, and J. Craninckx, "Wideband VCO with simultaneous switching of frequency band, active core, and varactor size," *IEEE J. Solid-State Circuits*, vol. 42, no. 7, pp. 1472–1480, Jul. 2007.

[10] N. T. Tchamov, S. S. Broussev, I. S. Uzunov, and K. K. Rantala, "Dual band LC VCO architecture with a fourth-order resonator," *IEEE Trans. Circuits Syst. II, Exp. Briefs*, vol. 54, no. 3, pp. 277–281, Mar. 2007.

[11] A. Buonomo and A. Lo Schiavog, "Analysis and design of dual-mode CMOS LC-VCOs," IEEE Trans. Circuits Syst. I, Reg. Papers, vol. 62, no. 7, pp. 1845–1853, Nov. 2015.

[12] A. Italia, C. Marco Ippolito, and G. Palmisano, "A 1-mW 1.13–1.9 GHz CMOS LC VCO using shunt-connected switched-coupled inductors," IEEE Trans. Circuits Syst. I, Reg. Papers, vol. 59, no. 6, pp. 1145–1155, Jun. 2012.

[13] B. Sadhu, J. Kim, and R. Harjani, "A CMOS 3.3-8.4 GHz wide tuning range, low phase noise LC VCO," *IEEE Custom Integrated Circuits Conf. (CICC)*, Sep. 2009.

[14] W. Deng, K. Okada, A Matsuzawa, "A 25MHz–6.44GHz LC-VCO using a 5-port inductor for multi-band frequency generation," *IEEE Radio Frequency IC Symposium*, Jun. 2011, pp. 1–4.

[15] S. Rong and H. C. Luong "Analysis and design of transformer-based dual-band VCO for software-defined radios," *IEEE Trans. on Microwave Theory and Techniques*, vol. 59, no. 3, pp. 449–462, Mar. 2012.

[16] A. El-Gouhary, N. M. Neihart, An analysis of phase noise in transformer-based dual-tank oscillators," *IEEE Trans. on Microwave Theory and Techniques*, vol. 61, no. 7, pp. 2098–2109, Jul. 2014.

[17] J. Yin and H. C. Luong, "A 57.5–90.1-GHz magnetically tuned multimode CMOS VCO," *IEEE J. Solid-State Circuits*, vol. 48, no. 8, pp. 1851–1861, Aug. 2013.

[18] U. Decanis, A. Ghilioni, E. Monaco, A. Mazzanti, and F. Svelto, "A low-noise quadrature VCO based on magnetically coupled resonators and a wideband frequency divider at millimeter waves," *IEEE J. Solid-State Circuits*, vol. 46, no. 12, pp. 2943–2955, Dec. 2011.

[19] G. Li and E. Afshari, "A distributed dual-band LC oscillator based on mode switching," *IEEE Trans. on Microwave Theory and Techniques*, vol. 59, no. 1, pp. 99–107, Jan. 2011.

[20] G. Li, Y. Tang, and E. Afshari, "A low phase-noise wide tuning-range oscillator based on resonant mode switching," *IEEE J. Solid-State Circuits*, vol. 47, no. 6, pp. 1295–1308, Jun. 2012.

[21] M. Taghivand, K. Aggarwal, and A. S. Y. Poon, "A 3.24-to-8.45 GHz low-phase-noise mode-switching oscillator," *in IEEE Int. Solid-State Circuits Conf. Dig Tech. Papers (ISSCC)*, Feb. 2014, pp. 368–370.

[22] F.-W. Kuo, R. Chen, K. Yen, H.-Y. Liao, C.-P. Jou, F.-L. Hsueh, M. Babaie, and R. B. Staszewski, "A 12mW all-digital PLL based on class-F DCO for 4G phones in 28nm CMOS," *Proc. of IEEE Symp. on VLSI Circuits (VLSI)*, sec. 9.4, pp. 1–2, June 2014.

[23] M. Babaie and R. B. Staszewski, "An ultra-low phase noise class-F_2 CMOS oscillator with 191dBc/Hz FOM and long term reliability," *IEEE J. Solid-State Circuits*, vol. 50, no. 3, pp. 679–692, Mar. 2015.

[24] M. Babaie and R. B. Staszewski, "A class-F CMOS oscillator," *IEEE J. Solid-State Circuits*, vol. 48, no. 12, pp. 3120–3133, Dec. 2013.

[25] A. Bevilacqua, F. P. Pavan, C. Sandner, A. Gerosa, and A. Neviani, "Transformer-based dual-mode voltage-controlled oscillators," *IEEE Trans. Circuits Syst. II, Exp. Briefs*, vol. 54, no. 4, pp. 293–297, Apr. 2007.

[26] A. Mazzanti and A. Bevilacqua, "On the Phase Noise Performance of Transformer-Based CMOS Differential-Pair Harmonic Oscillators," *IEEE Trans. Circuits Syst. I, Reg. Papers*, vol. 62, no. 9, pp. 293–297, Sep. 2015.

[27] P. Andreani et al., "A study of phase noise in Colpitts and LC-tank CMOS oscillators," *IEEE J. Solid-State Circuits*, vol. 40, no. 5, pp. 1107–1118, May 2005.

[28] A. Hajimiri and T. H. Lee, "A general theory of phase noise in electrical oscillators," *IEEE J. Solid-State Circuits*, vol. 33, no. 2, pp. 179–194, Feb. 1998.

[29] D. Murphy, J. J. Rael, and A. A. Abidi "Phase noise in LC oscillators: A phasor-based analysis of a general result and of loaded Q," *IEEE Trans. Circuits Syst. I, Reg. Papers*, vol. 57, no. 6, pp. 1187–1203, June 2010.

[30] D. Chowdhury, L. Ye, E. Alon, and A. M. Niknejad, "An efficient mixed-signal 2.4-GHz polar power amplifier in 65-nm CMOS technology," *IEEE J. Solid-State Circuits*, vol. 46, no. 8, pp. 1796–1809, Aug. 2011.

[31] B. Soltanian and P. Kinget, "A low phase noise quadrature LC VCO using capacitive common-source coupling," in *IEEE European Solid-State Circuits Conference (ESSCIRC)*, 2006, pp. 436–439.

[32] Y. Wu, M. Shahmohammadi, Y. Chen, P. Lu, and R. B. Staszewski, "A 3.5–6.8-GHz wide-bandwidth DTC-assisted fractional-N all-digital PLL with a MASH $\Delta\Sigma$-TDC for low in-band phase noise," *IEEE Journal of Solid-State Circuits (JSSC)*, vol. 52, no. 7, pp. 1885–1903, Jul. 2017.

[33] M. Shahmohammadi, M. Babaie, and R. B. Staszewski, "Tuning range extension of a transformer-based oscillator through common-mode Colpitts resonance," *IEEE Trans. on Circuits and Systems I (TCAS-I)*, vol. 64, no. 4, pp. 836–846, Apr. 2017.

[34] M. Shahmohammadi, M. Babaie, and R. B. Staszewski, "Radio frequency oscillator," *US Patent 2017/0324378*, published 9 Nov. 2017.

8

A Study of RF Oscillator Reliability
in Nanoscale CMOS

In this chapter, we investigate the nature of oxide breakdown and stress-related degradation mechanisms in MOS transistors. The MOS breakdown time is quantified based on exponential-law and defect-generation models versus the oxide-thickness, gate area, temperature, and voltage stress at a given cumulative failure. As a consequence, a design guide is presented to estimate the time-dependent dielectric breakdown of any analog circuit. Based on reliability analysis, the lifetime of class-F_3 oscillator of Chapter 3 is evaluated for both thin- and thick-oxide options in TSMC 65-nm CMOS process as a case study. The long-term reliability is also investigated for class-F_2 oscillator introduced in Chapter 4.

8.1 Introduction

To keep on implementing increasingly complex functions while reducing the overall solution costs, scaling of CMOS transistors is inevitable. As circuits are growing denser, all of the physical dimensions of the transistors must be reduced correspondingly. The SiO_2 oxide-layer thickness reduction is accompanied by migrating to smaller supply voltages. This is to maintain the electric field strength across the oxide in order to prevent the device performance degradation due to the time-dependent dielectric breakdown (TDDB) [1]. Although digital circuits have fared well, analog designers face additional difficulties with the transistor scaling. The supply voltage V_{DD} is reduced while RF and analog circuits must maintain their dynamic range, noise performance, and output power. For example, the oscillator phase noise performance and power amplifier (PA) output power degrade by 6 dB/octave with reduction of their supply voltage [2, 3]. On the other hand, LC-tank oscillators and PAs usually operate at a voltage swing in excess of the nominal supply voltage. This causes potential reliability issues due to the large electric field across the gate oxide. Consequently, analog designers must consider the reliability of the circuit while trying to maximize the voltage swing to reach better output power, dynamic range, or noise performance.

In the classical view, the reliability must be qualified at the technology level and guaranteed by the manufacturer. However, this perception is no longer valid. The circuit reliability has become highly circuit-dependent in the advanced CMOS technologies. The designers have to improve the reliability margins by adapting their approach and taking into account the impact of failure at the circuit level. In this chapter, we investigate the nature of oxide breakdown and stress-related degradation mechanisms in MOS transistors. The maximum gate-oxide voltage of a MOS transistor is quantified versus the oxide thickness, gate area, and temperature for different cumulative failure rates and operating times. We exemplify the oxide breakdown reliability calculations in class-F RF oscillators of Chapters 3 and 4.

8.2 Gate-Oxide Breakdown

Gate-oxide breakdown leads to a catastrophic and permanent failure in MOS devices. The breakdown is accompanied by a sudden discontinuous increase in the oxide conductance and the gate current noise. Breakdown is a gradually increasing phenomenon and realized by defects such as electron traps in the oxide structure. The rate of defect generation is almost proportional to the gate current density. As a consequence, the transistors with a smaller channel length are more vulnerable. The gate current is due to Fowler–Nordheim (FN) tunneling for thick-oxide devices at a gate voltage V_g above 3 V, while it is due to a direct quantum-mechanical tunneling (DT) for thin oxides ($t_{ox} \leq$ 3-nm) at voltages below 3 V [4]. These gate currents trigger "impact ionization", "anode hole injection", and "trap creation" phenomena to generate defects in the oxide structure. Then, the probable breakdown will occur at a critical trap density by a conduction path via these generated traps. Consequently, the oxide breakdown failure is a time-dependent and statistically distributed phenomenon. It is well known that the oxide breakdown can be described by the Weibull distribution [1]:

$$F\left(T_{BD}\right) = 1 - e^{-\left(\frac{T_{BD}}{\eta}\right)^{\beta}} \tag{8.1}$$

where F is a cumulative failure probability and T_{BD} is a random variable for time-to-breakdown. η is a characteristic time-to-breakdown at 63.2% failure probability and β is a Weibull shape slope.

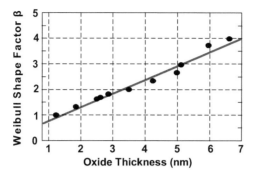

Figure 8.1 Weibull slope versus gate-oxide thickness.

8.2.1 Weibull Slope

Figure 8.1 shows measured Weibull slopes obtained from literature [5] versus oxide thickness ranging from 1.25 to 7 nm at a temperature of 140°C. The solid curve is a linear fit of an analytical cell-based model and expressed by

$$\beta = \frac{t_{int} + t_{ox}}{\alpha_0}, \tag{8.2}$$

where α_0 is the defect size that is found to be 1.83 nm in [5] and t_{int} is the interface thickness, which was reported 0.37 nm in [5]. Equation (8.2) indicates that Weibull slope decreases with the technology scaling. Suppose the η value is the same for both thin- and thick-oxide devices. Although both devices reach the cumulative failure rate of 63% at the same time, the early failure rate of a thin-oxide transistor will be much larger than that of the thick-oxide device due to its lower Weibull slope. Consequently, thin-oxide devices can only tolerate lower electric field strengths for the certain failure rate in a given operating time. It is concluded in [1] that Weibull slope is independent over a wide range of stress voltage, temperature, and polarity.

8.2.2 η Estimation for Different Oxide Thicknesses

Figure 8.2 shows η versus gate voltage for different oxide thicknesses. The data points (open/solid squares, triangles, and circles) are extracted from literature [1, 4, 9] and scaled to 140°C and an area of 10^3 μm^2. The TDDB reliability is usually estimated by means of voltage and temperature acceleration models from results acquired at relatively short measurement times to the required product lifetime of 10 years or more. Such a scaling may span several decades and magnify inaccuracies if the model is not correct or the breakdown mechanism changes along the voltage scaling. Until now, at least

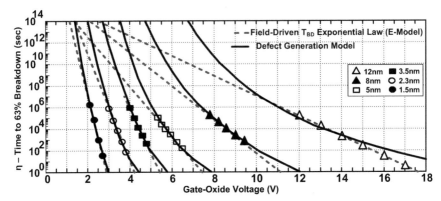

Figure 8.2 Comparison of characteristic time-to-breakdown η versus gate voltage in NMOS inversion for different gate-oxide thicknesses from 1.5 to 12 nm. The solid lines represent the result of defect generation model as described in [7]. The dashed lines are from the least-square fit using the E-model, as described in [9]. The data points (open/solid squares, triangles, and circles) are extracted from literature and scaled to 140°C with an area of 10^3 μm^2.

five voltage acceleration models have been proposed: E-model, 1/E model, power-law model, 1/V model, and physics-based model. Not surprisingly, it is confusing and practically impossible to decide which model should be used in TDDB calculations.

The field-driven E-model refers to the experimental observation that T_{BD} data can be characterized by $exp(\gamma \cdot E_{ox})$, where the electric field E_{ox} is considered as a variable in TDDB process [9]. The η variations based on E-model curve fitting are illustrated by dashed lines in Figure 8.2 for different oxide thicknesses. The model can be safely ruled out for both thin and thick oxide at $E_{ox} \geq 7$ MV/cm. However, an extrapolation to lower fields would result in a cross-over between the dashed lines meaning thick-oxide devices are less reliable than thin-oxide ones at voltages below 1.5 V, which is contradictory to the fundamental physics. Consequently, this model is not accurate at $E_{ox} \leq 7$ MV/cm. Nevertheless, it is possible to use E-model as a conservative projection. Hence, E-model estimation has been added in Figure 8.2 as the worst case in T_{BD} prediction [7].

The anode injection model (1/E model) characterizes T_{BD} based on the FN tunneling current. However, the direct tunneling is a dominant phenomenon at $V_g \leq 4$ V. Hence, 1/E model is not applicable and leads to optimistic results at lower voltages [8].

A more realistic projection is the physics-based breakdown model, which considers both tunneling current and defect generation phenomena.

This model is consistent with many measurement results up to the 8-nm oxide thickness. The discrepancy at thicker oxides originates in "band-to-band ionization", which plays an important role in very thick oxides stressed at relatively high voltages [7]. The η extrapolation based on "defect-generation" model is also added by solid lines in Figure 8.2 for different oxide thicknesses.

8.2.3 Area and Temperature Dependence of T_{BD}

The failure of an entire IC chip is defined by the first failure of a single device. From elementary statistics, if the failure probability of a unit is F_{A1}, then the failure probability of a circuit comprising N independent units is given by

$$F_{A2} = 1 - (1 - F_{A1})^N .$$

(8.3)

By substituting (8.1) into (8.3) and carrying out lengthy algebra, the area scaling equation of η is obtained by

$$\eta_2 = \eta_1 \left(\frac{A_1}{A_2} \right)^{\frac{1}{\beta}} ,$$

(8.4)

where A_1 and A_2 correspond to two different areas of the oxide. This expression shows that the characteristic breakdown time, η, increases with reducing the oxide area. The area scaling is a strong function of the Weibull slope and, consequently, thinner oxides are more sensitive. For example, in 10^{-4} to $10 \ mm^2$ area scaling, T_{BD} lifetime drops with 3–4 orders of magnitude for a 4-nm oxide. However, the T_{BD} reduction would be just a factor of 2 for an 11-nm oxide.

The following equation expresses the dependency of η on the oxide junction temperature [6]. As expected, higher temperatures accelerate the TDDB process.

$$\frac{\eta_2}{\eta_1} = e^{\frac{E_a}{K_B} \left(\frac{1}{T_2} - \frac{1}{T_1} \right)} ,$$

(8.5)

where K_B is the Boltzmann's constant and E_a is the thermal activation energy that is about $1 \ eV$ and changes by a small amount with the gate voltage [6].

8.2.4 Principle of Extrapolation to a Specified Condition

Figure 8.2 shows the extrapolated η for physics-based and E models for an area $10^3 \ \mu m^2$ at $140°C$. The main question arises: How can one predict

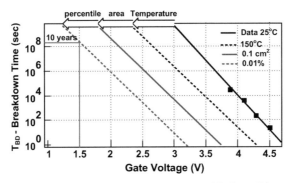

Figure 8.3   Extrapolation steps to the specified condition.

TDDB lifetime of a circuit for a given voltage and cumulative failure values based on this figure? The principle of T_{BD} extrapolation to a specified condition is illustrated in Figure 8.3. As a first step, one of the curves in Figure 8.2 is chosen based on the technology oxide thickness. This curve will shift through (8.5) if the operating temperature deviates from the reference value $140°C$. Then, the area scaling is applied to the graph using (8.4). The curve is scaled once more to the desired cumulative failure by (8.1). Finally, the corresponding T_{BD} can be calculated from the obtained curve (blue dashed-line in Figure 8.3) for any gate voltages.

8.3 Hot Carrier Degradation

Hot carriers (HC) are holes or electrons that are accelerated to high energies by an electric field caused by a large drain–source voltage. Certain percentage of the hot carriers collide with the lattice and create electron–hole pairs. Furthermore, if the hot charge carriers have a kinetic energy larger than the silicon-oxide barrier height, some of them will dominate the barrier and flow toward the insulator.

Unlike the gate-oxide breakdown, this phenomenon is not inherently catastrophic. Instead, it can cause a gradual performance degradation during the operating lifetime. These traps can shift the threshold voltage and reduce the conducting carrier mobility. Consequently, the drain current, channel resistance, and transconductance gain of MOS transistor decrease and degrade performance of RF circuits, such as oscillators. First, one needs to choose larger gm-devices to compensate the oscillator loop gain reduction due to HC. It means the active device injects more noise into the tank, resulting in an increase of the oscillator's effective noise factor. Furthermore, the circuit phase noise performance gradually degrades due to reduction of drain current and thus the oscillation voltage swing. Hence, an additional mechanism should sense the oscillation amplitude in order to increase the drain current of active devices by adjusting their bias voltage. Second, a combination of a large parasitic capacitance of the tail current transistor and a smaller channel resistance of the gm-device could provide a discharge path between the tank and ground. It would drop the equivalent quality factor of the tank resulting in phase noise degradation. Third, hot carrier stress also increases the $1/f$ noise of the MOS transistor, which is translated to a larger $1/f^3$ oscillator phase noise corner.

However, the hot carrier degradation mainly occurs when the drain current and drain–source voltage are substantial at the same time [3]. Hence, the hot carrier degradation would be negligible if the channel current was low when the drain–source voltage was high and vice versa. This condition naturally occurs in oscillators and switching power amplifiers. Consequently, the RF oscillators are not inherently vulnerable to the hot carrier degradation.

8.4 Negative Bias Temperature Instability

The negative bias temperature instability (NBTI) occurs when a negative gate–source voltage V_{gs} is applied causing an increase in the absolute threshold voltage, a degradation of the mobility, drain current, and transconductance. PMOS devices are more vulnerable to NBTI. Although NMOS devices can be damaged in an NBTI stress, the damage occurs at negative V_{gs} where NMOS devices are not active. Consequently, NMOS devices are thus suggested for long operating time applications such as infrastructure basestations and satellite communications.

8.5 Reliability of Class-F$_3$ Oscillators

8.5.1 Class-F$_3$ Oscillators

Figure 8.4(a) shows a schematic of the class-F$_3$ oscillator in Chapter 3, which was realized in TSMC 65-nm CMOS technology. It exhibits a pseudo-square-wave across the tank to desensitize the oscillator phase noise to the circuit noise. Figure 8.4(b) illustrates the oscillation waveforms. As can be seen, gate–drain voltage increases to 3.2 V at 1.2-V supply voltage due to the passive voltage gain of the transformer. $M_{1/2}$ dimensions should be (48-μm/0.28-μm) and (12-μm/65-nm) for thick (5 nm) and thin (2.3 nm)

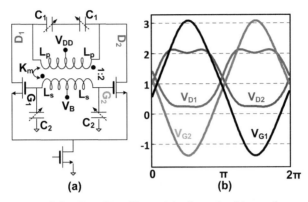

Figure 8.4   Class-F oscillator: (a) schematic; (b) waveforms.

oxide options, respectively, to guarantee safe oscillator start-up in all process corners. Let us investigate the oscillator's lifetime at 0.01% cumulative failure rate and 140°C for both transistors. From Figure 8.5, the η values are obtained as

$$\eta_{thin}\left(10^3 \mu m^2\right) = 10^5 s, \quad \eta_{thick}\left(10^3 \mu m^2\right) = 10^{10} s. \tag{8.6}$$

The conservative E-model is used for the T_{BD} calculations. Weibull slope should be determined in order to extrapolate η values to the desired area. Based on (8.2),

$$\beta_{thin} = 1.5, \quad \beta_{thick} = 3. \tag{8.7}$$

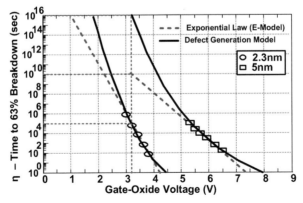

Figure 8.5   Class-F$_3$ oscillator lifetime estimation due to TDDB for thin- and thick-oxide transistors.

The next step is to apply the area scaling factor to η by (8.4).

$$\eta_{thin}\left(1.6\mu m^2\right) = 7\cdot 10^6 \text{s}, \quad \eta_{thick}\left(27\mu\text{m}^2\right) = 3\cdot 10^{10}\text{s}. \quad (8.8)$$

Finally, the lifetime can be estimated by substituting the calculated parameters in (8.4):

$$T_{BD}\left(thin\right) = 4\ hours, \quad T_{BD}\left(thick\right) = 40\ years \quad (8.9)$$

The oscillator lifetime drops dramatically by 4 orders of magnitude just by replacing thick-oxide gm-devices with thin-oxide ones. Consequently, a thick-oxide device must be used in the class-F_3 oscillator to satisfy the required T_{BD} and failure rate but at a cost of more parasitic capacitance and lower tuning range.

8.5.2 Class-F_2 Oscillators

The HCI degradation would occur when the drain current, I_{DS}, and drain–source voltage, V_{DS}, are large at the same time. Thanks to the transformer's voltage gain, in class-F_2 oscillator, V_{DD} is low enough such that V_{DS} of its gm-devices can be much less than the standard voltage of thick-oxide transistors (2.5 V) when they operate in on-state (see Figure 4.11). Consequently, this oscillator is not inherently vulnerable to HCI. However, the large oscillation swing applies a strong electric field across the gate oxide of gm-devices (V_{DG}, V_{GS}), which can potentially reduce the long-term reliability of the oscillator due to TDDB.

The oxide breakdown stems from defects, such as electron traps, in the oxide structure. The rate of defect generation is almost proportional to the gate-oxide electric field and its leakage current density. We can re-write Equation (8.1) as

$$\eta = T_{BD}\left(-ln\left(1-F\right)\right)^{-1/\beta}. \quad (8.10)$$

It is shown in [1] that η for a given circuit with arbitrary characteristics (A_{ox}, V_{ox}, and T_{ox}) can be extrapolated from the reference data (x_{ref}) by

$$\eta = \eta_{ref}\left(\frac{V_{ox}}{V_{ref}}\right)^{-n} e^{\frac{E_a}{K}\left(\frac{1}{T_{ox}}-\frac{1}{T_{ref}}\right)}\left(\frac{A_{ox}}{A_{ref}}\right)^{-1/\beta}. \quad (8.11)$$

where n is voltage acceleration factor.

We can apply the above procedure to our class-F_2 oscillator to determine its T_{BD}. Figure 8.6(a) shows the measured F versus T_{BD} for 14 samples

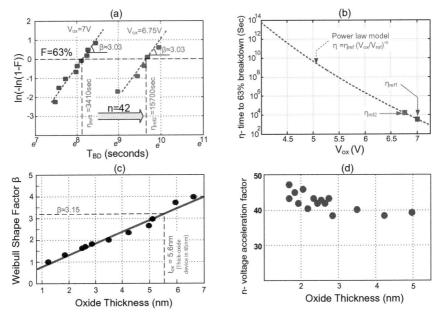

Figure 8.6 (a) Measured cumulative failure rate F versus breakdown time T_{BD} for 14 samples of a thick-oxide transistor (176 μm/0.28 μm) at room temperature, (b) the projected η value versus different gate-oxide stress voltage based on the measured η_{ref}, (c) Weibull slope versus gate-oxide thickness extracted from measurement results in [1], and (d) voltage acceleration versus gate-oxide thickness extracted from measurement results in [10].

of the thick-oxide transistor (176 μm/0.28 μm) at room temperature when a large voltage (6.75 and 7 V) is applied across the gate. The data points are easily mapped to a Weibull distribution curve as indicated by the dashed line. The cross-over of these curves at $F = 63.2\%$ specifies the reference η values (η_{ref}). The voltage acceleration ratio n is calculated by applying η_{ref} values and their related V_{ox} in (8.11). Furthermore, the slope of the curves determines β. Consequently, the estimated n and β values are, respectively, 42 and 3 for the thick-oxide devices ($t_{ox} = 5.6$ nm) in 65-nm CMOS, which are close to extracted measured numbers from literature, as shown in Figure 8.6(c,d) [1, 10]. The E_a is ~0.55 eV and independent from the oxide thickness and temperature [11]. Consequently, the given oscillator η can be estimated by substituting the measured reference and technology parameters and circuit characteristics (A_{ox}, V_{ox}, and T_{ox}) in (8.11). Finally, T_{BD} is calculated by substituting the estimated η and the desired F in (8.10).

The lifetime estimation of our circuit as a function of V_{ox} is plotted in Figure 8.7 for various F, T_{ox}, and A_{ox}. The plots indicate that the maximum voltage across the oxide for $M_{1,2}$ transistors should be <4.4 V to ensure <0.01% failure during 10 years at 125°C. The max V_{ox} could be increased if higher failure rate or lower max operating temperature are accepted. The maximum dc voltage is thus established across the gate oxide. However, the actual nature of stress in RF oscillators is not dc but an ac voltage $V_{ox}(\omega_0 t)$. Consequently, it is instructive to compare the static max V_{ox} with the actual operation when η changes over the period of the resonant frequency. Hence, the "effective" η is calculated as

$$\frac{1}{\eta_{\text{eff}}} = \frac{1}{2\pi} \int_0^{2\pi} \frac{1}{\eta\left(V_{ox}\left(\omega_0 t\right)\right)} \, d\left(\omega_0 t\right), \qquad (8.12)$$

where $\eta(V_{ox}(\omega_0 t))$ is given by (8.11) and can be expediently simplified to $\eta = B \cdot (V_{ox}(\omega_0 t))^{-n}$.

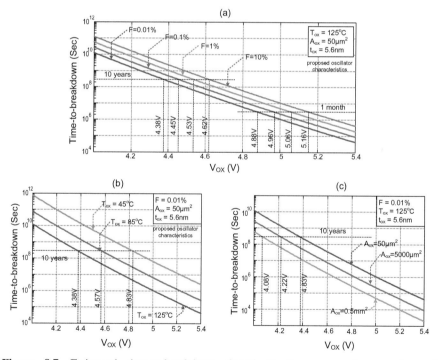

Figure 8.7 Estimated time-to-breakdown (based on the measured parameters of Figure 8.6(a)) of thick-oxide transistors in 65-nm CMOS versus maximum gate-oxide stress voltage for different (a) cumulative failure rates, (b) temperatures, and (c) gate-oxide areas.

Starting with the application's desired operating time (i.e., T_{BD}) at a given failure rate (F) in a given technology (i.e., β), the parameter η_{eff} is first established as per (8.10) and is identical for dc and ac operations. For a dc operation, $\eta = B \cdot (V_{dc})^{-n}$ and (8.12) results in

$$V_{dc} = \left(\frac{B}{\eta_{eff}} \right)^{1/n}.$$ (8.13)

However, for an ac operation,

$$\frac{1}{\eta_{eff}} = \frac{1}{2\pi} \int_0^{2\pi} \frac{1}{B \left(0.5 V_{ac,max} \left(1 + sin \left(\omega_0 t \right) \right) \right)^{-n}} d \left(\omega_0 t \right).$$ (8.14)

Solving this integral for the voltage acceleration factor n of 42 for the 65-nm CMOS thick-oxide devices,

$$V_{ac,max} = \left(\frac{11.5 \cdot B}{\eta_{eff}} \right)^{1/n}.$$ (8.15)

Consequently, the ac to dc maximum tolerable stress voltage ratio $(V_{ac,max}/V_{dc})$ will be $(11.5)^{1/n} \approx 1.06$. We strongly emphasize that there are no significant differences in max V_{ox} at ac-peak and dc conditions due to the sharp slope of $T_{BD} - V_{ox}$ curves in Figure 8.7. As shown by integrating the voltage-dependent $\eta(V_{ox})$ over the full oscillation cycle, the peak magnitude of the V_{ox} sine wave can be just 6% higher than what is determined for a fixed dc V_{ox}. Consequently, the slightly lower pessimistic value of V_{ox} in the dc condition could be used as an extra margin.

8.6 Conclusion

Time-dependent dielectric breakdown (TDDB), hot carrier degradation (HCI), and negative bias temperature instability (NBTI) mechanisms were investigated for a MOS transistor. The exponential-law and defect-generation models quantified the MOS breakdown lifetime versus the oxide-thickness, gate area, temperature, gate voltage, and cumulative failure rate. A design guide is presented to estimate TDDB of any analog circuit. Based on reliability analysis, a huge 4-decade lifetime difference exists between thin (2.3-nm) and thick (5-nm) oxide devices in the class-F_3 oscillator. The reliability process is highly circuit-dependent in the advanced CMOS technologies (oxide thickness $\leq$ 3 nm) and analog/RF engineers have to consider the reliability issues in the design cycle.

References

[1] E. Y. Wu, et al., "CMOS scaling beyond the 100-nm node with silicon-dioxide-based gate dielectrics," *IBM Journal of Research and Development*, vol. 46, no. 2/3, pp. 287–297, Mar./May 2002.

[2] M. Babaie and R. B. Staszewski, "Third-harmonic injection technique applied to a 5.87- to 7.56 GHz 65nm class-F oscillator with 192dBc/Hz FoM," *Proc. of IEEE Solid-State Circuits Conf.*, pp. 348–349, Feb. 2013.

[3] I. Aoki, et al., "A fully integrated quad-band GSM/GPRS CMOS power amplifier," *IEEE Journal of Solid-State Circuits*, vol. 43, no. 12, pp. 2747–2758, Dec. 2008.

[4] J. H. Stathis, "Physical and predictive models of ultrathin oxide reliability in CMOS devices and circuits," *IEEE Transactions on Device and Materials Reliability*, vol. 1, no. 1, pp. 43–59, Mar. 2001.

[5] E. Y. Wu, J. Sune, and W. Lai, "On the Weibull Shape Factor of Intrinsic Breakdown of Dielectric Films and Its Accurate Experimental Determination – Part II: Experimental Results and the Effects of Stress Conditions," *IEEE Transactions on Electron Devices*, vol. 49, no. 12, pp. 2141–2150, Dec. 2002.

[6] A. M. Yassine, et al., "Time dependent breakdown of ultrathin gate oxide" *IEEE Transactions on Electron Devices*, vol. 47, no. 7, pp. 1416–1420, Jul. 2000.

[7] E. Y. Wu, et al., "Comprehensive physics-based breakdown model for reliability assessment of oxides with thickness ranging from 1 nm up to 12 nm," *Proceedings of the 47th Annual International Reliability Physics Symposium (IRPS)*, pp. 708–717, Montreal 2009.

[8] I. C. Chen, et al., "Electrical breakdown in thin gate and tunneling oxides," *IEEE Journal of Solid-State Circuits*, vol. 20, no. 1, pp. 333–342, Feb. 1985.

[9] A. Berman, "Time-zero dielectric reliability test by a ramp method," *in IEEE Proc. International Reliability Physics Symp.*, 1981, pp. 204–209.

[10] E. Wu, J. Aitken, E. Nowak, A. Vayshenker, P. Varekamp, G. Hueckel, J. McKenna, D. Harmon, L. K. Han, C. Montrose, and R. Dufresne, "Voltage-dependent voltage acceleration of oxide breakdown for ultra-thin oxides," *IEEE International Electron Devices Meeting (IEDM)*, 2000, pp. 541–544.

[11] E. Wu, J. McKenna, W. Lai, E. Nowak, and A. Vayshenker, "The effect of change of voltage acceleration on temperature activation of oxide breakdown for ultrathin oxides," *IEEE Electron Device Letters*, vol. 22, no. 12, pp. 603–605, Dec. 2001.

Index

About the Authors

Masoud Babaie received the B.Sc. degree (Hons.) in electrical engineering from the Amirkabir University of Technology, Tehran, Iran, the M.Sc. degree in electrical engineering from the Sharif University of Technology, Tehran, and the Ph.D. degree (cum laude) in electrical engineering from the Delft University of Technology, Delft, The Netherlands, in 2004, 2006, and 2016, respectively.

In 2006, he joined the Kavoshcom Research and Development Group, Tehran, where he was involved in designing wireless communication systems. From 2009 to 2011, he was a CTO of that company. He was consulting for RF group of TSMC, Hsinchu, Taiwan, in 2013–2015, designing 28-nm All-Digital PLL and Bluetooth Low Energy transceiver chips. From 2014 to 2015, he was a Visiting Scholar Researcher with the Berkeley Wireless Research Center, Berkeley, CA, USA. In 2016, he joined the Delft University of Technology, Delft, The Netherlands, as an Assistant Professor. His current research interests include RF/millimeter-wave integrated circuits and systems for wireless communications, and cryogenic electronics for quantum computation.

Dr. Babaie has been a Committee Member of Student Research Preview (SRP) of the IEEE International Solid-State Circuits Conference (ISSCC), since 2017. He was a recipient of the 2015–2016 IEEE Solid-State Circuits Society Pre-Doctoral Achievement Award. In 2019, he received the Veni award from The Netherlands Organization for Scientific Research (NWO).

Mina Shahmohammadi received the B.Sc. degree in communication systems from the Amirkabir University of Technology, Tehran, Iran, in 2005, the M.Sc. degree in electronics from the University of Tehran, Tehran, in 2007, and the Ph.D. degree from Electronics Research Laboratory, Delft University of Technology (TU Delft), Delft, The Netherlands, in 2016, with a focus on wide tuning range and low flicker noise RF-CMOS oscillators.

From 2007 to 2011, she was with Rezvan Engineering Company, Tehran, as an Analog Designer. She was a Research Assistant with the Electronic

Instrumentation Laboratory, TU Delft, from 2011 to 2013, where she was involved in resistor-based temperature sensors. She is currently an Analog Designer with Catena B.V., Delft. Her current research interests include analog and RF integrated circuits design.

Robert Bogdan Staszewski was born in Bialystok, Poland. He received the B.Sc. (*summa cum laude*), M.Sc., and Ph.D. degrees in electrical engineering from the University of Texas at Dallas, Richardson, TX, USA, in 1991, 1992, and 2002, respectively.

From 1991 to 1995, he was with Alcatel Network Systems in Richardson, TX, USA, involved in SONET cross-connect systems for fiber optics communications. He joined Texas Instruments Incorporated, Dallas, TX, USA, in 1995 where he was elected Distinguished Member of Technical Staff (limited to 2% of technical staff). From 1995 to 1999, he was engaged in advanced CMOS read channel development for hard disk drives. In 1999, he co-started the Digital RF Processor (DRP) group within Texas Instruments with a mission to invent new digitally intensive approaches to traditional RF functions for integrated radios in deeply-scaled CMOS technology. He was appointed as a CTO of the DRP group from 2007 to 2009. In 2009, he joined the Delft University of Technology, Delft, The Netherlands, where currently he holds a guest appointment of Full Professor (*Antoni van Leeuwenhoek Hoogleraar*). Since 2014, he has been a Full Professor with the University College Dublin (UCD), Dublin, Ireland. He has authored or co-authored four books, five book chapters, 260 journal and conference publications, and holds 180 issued US patents. His research interests include nanoscale CMOS architectures and circuits for frequency synthesizers, transmitters and receivers.

Prof. Staszewski has been a TPC member of ISSCC, RFIC, ESSCIRC, ISCAS and RFIT. He is an upcoming TPC Chair of 2019 ESSCIRC in Krakow, Poland. He was a recipient of the 2012 IEEE Circuits and Systems Industrial Pioneer Award.